Milk: Bioactive Components and Role in Human Nutrition

Special Issue Editor
Alessandra Durazzo

MDPI • Basel • Beijing • Wuhan • Barcelona • Belgrade

MDPI

Special Issue Editor
Alessandra Durazzo
Consiglio per la ricerca in agricoltura e l'analisi dell'economia agraria Centro di ricerca
CREA—Alimenti e Nutrizione
Italy

Editorial Office
MDPI AG
St. Alban-Anlage 66
Basel, Switzerland

This edition is a reprint of the Special Issue published online in the open access journal *Beverages* (ISSN 2306-5710) in 2017 (available at: http://www.mdpi.com/journal/beverages/special_issues/ bioactive_components).

For citation purposes, cite each article independently as indicated on the article page online and as indicated below:

Lastname, F.M., and F.M. Lastname. Year. Article title. *Journal Name* Article number: page range.

First Edition 2018

ISBN 978-3-03842-723-0 (Pbk)
ISBN 978-3-03842-724-7 (PDF)

Table of Contents

About the Special Issue Editor

Alessandra Durazzo was awarded a Master's degree in Chemistry and Pharmaceutical Technology cum laude in 2003, a PhD in Horticulture in 2010, the National Scientific Habilitation as Associate Professor (13/B5 Commodity Science) in 2013. Since 2005, Researcher at the Consiglio per la ricerca in agricoltura e l'analisi dell'economia agraria—Centro di ricerca CREA—Alimenti e Nutrizione. The core of her research is the study of chemical, nutritional and bioactive components of food, with particular regard to the wide spectrum of substances classes and their nutraceutical features. For several years, she was involved in national and international research projects on evaluation of several factors (agronomic practices, processing, etc.) that affect food quality, the levels of bioactive molecules and the total antioxidant properties, as well as on their possible impact on the biological role played by bioactive components in human physiology. Her research activities are addressed also towards the development, management and updating of the Food Composition Database, as well as Bioactive Compounds and Food Supplements databases; particular attention was given towards the harmonization of analytical procedures and classification and codification of food preparation and food supplements.

beverages

MDPI

Editorial

Milk: Bioactive Components and Role in Human Nutrition

Alessandra Durazzo

Consiglio per la ricerca in agricoltura e l'analisi dell'economia agraria—Centro di ricerca CREA—Alimenti e Nutrizione, Via Ardeatina 546, 00178 Rome, Italy; alessandra.durazzo@crea.gov.it; Tel.: +39-06-514-94430; Fax: +39-06-514-94550

Academic Editor: Edgar Chambers IV
Received: 28 November 2017; Accepted: 14 December 2017; Published: 19 December 2017

In the current Special Issue, numerous and different aspects related to milk, an important component of a well-balanced diet, are presented.

Several of the scientists that contributed to this Special Issue investigated and tested the effectiveness of actions targeting the promotion of milk, and an increase in general daily milk consumption, especially for children: an important goal to achieve [1–4]. Gennaro et al. [3] provide an updated picture of communication strategies developed to improve healthy dietary habits in schoolchildren, with a focus on the importance of milk consumption. An example of this strategy is given by Emerson et al. [1]: small prizes increased plain milk and vegetable selection by elementary schoolchildren without adversely affecting total milk purchase. Two [2,4] of these studies are addressed towards consumer reactions relating to interventions promoting use of 1% low-fat milk.

It is worth mentioning the work of Lucarini [5] on the bioactive peptides of milk, from encrypted sequences to healthy applications; the author underlines how the exploitation of chemistry, bioavailability and the biochemical properties of bioactive peptides represent a key tool for nutraceutical and functional foods, also in the areas of circular bioeconomy and biorefinery [5]. In this context, the review by Vincenzetti et al. [6] focuses on the role of proteins and some bioactive peptides on nutritional quality, as well as the potential beneficial properties of donkey milk.

The microbiological, nutritional and sensory profile of raw and heat-treated milk are described in the work of Melini et al. [7], in order to evaluate the real risks and benefits of its consumption. Then, Gambelli [8] provides an updated picture of methodologies for the assessment of lactose. Some studies on the potential benefits of some milk components [9,10], as well as a consideration from a nutritional point of view of organic vs. conventional milk with regard to fat-soluble vitamins and iodine content [11], are also discussed.

I would like to acknowledge the efforts of the authors of the publications in this Special Issue. Their contributions will help to improve the understanding and promotion the key role of milk in nutrition.

References

1. Emerson, M.; Hudgens, M.; Barnes, A.; Hiller, E.; Robison, D.; Kipp, R.; Bradshaw, U.; Siegel, R. Small Prizes Increased Plain Milk and Vegetable Selection by Elementary School Children without Adversely Affecting Total Milk Purchase. *Beverages* **2017**, *3*, 14. [CrossRef]
2. Finnell, K.J.; John, R. Research to Understand Milk Consumption Behaviors in a Food-Insecure Low-Income SNAP Population in the US. *Beverages* **2017**, *3*, 46. [CrossRef]
3. Gennaro, L.; Durazzo, A.; Berni Canani, S.; Maccati, F.; Lupotto, E. Communication Strategies to Improve Healthy Food Consumption among Schoolchildren: Focus on Milk. *Beverages* **2017**, *3*, 32. [CrossRef]
4. John, R.; Finnell, K.J.; Kerby, D.S.; Owen, J.; Hansen, K. Reactions to a Low-Fat Milk Social Media Intervention in the US: The Choose 1% Milk Campaign. *Beverages* **2017**, *3*, 47. [CrossRef]

5. Lucarini, M. Bioactive Peptides in Milk: From Encrypted Sequences to Nutraceutical Aspects. *Beverages* **2017**, *3*, 41. [CrossRef]

6. Vincenzetti, S.; Pucciarelli, S.; Polzonetti, V.; Polidori, P. Role of Proteins and of Some Bioactive Peptides on the Nutritional Quality of Donkey Milk and Their Impact on Human Health. *Beverages* **2017**, *3*, 34. [CrossRef]

7. Melini, F.; Melini, V.; Luziatelli, F.; Ruzzi, M. Raw and Heat-Treated Milk: From Public Health Risks to Nutritional Quality. *Beverages* **2017**, *3*, 54. [CrossRef]

8. Gambelli, L. Milk and Its Sugar-Lactose: A Picture of Evaluation Methodologies. *Beverages* **2017**, *3*, 35. [CrossRef]

9. Gupta, C.; Prakash, D. Therapeutic Potential of Milk Whey. *Beverages* **2017**, *3*, 31. [CrossRef]

10. Norris, G.H.; Porter, C.M.; Jiang, C.; Blesso, C.N. Dietary Milk Sphingomyelin Reduces Systemic Inflammation in Diet-Induced Obese Mice and Inhibits LPS Activity in Macrophages. *Beverages* **2017**, *3*, 37. [CrossRef]

11. Manzi, P.; Durazzo, A. Organic vs. Conventional Milk: Some Considerations on Fat-Soluble Vitamins and Iodine Content. *Beverages* **2017**, *3*, 39. [CrossRef]

beverages

MDPI

Review

Communication Strategies to Improve Healthy Food Consumption among Schoolchildren: Focus on Milk

Laura Gennaro *, Alessandra Durazzo, Sibilla Berni Canani , Fabrizia Maccati and Elisabetta Lupotto

Consiglio per la Ricerca in Agricoltura e l'Analisi dell'Economia Agraria,
Centro di Ricerca CREA-Alimenti e Nutrizione, Via Ardeatina 546, 00178 Roma, Italy;
alessandra.durazzo@crea.gov.it (A.D.); sibilla.bernicanani@crea.gov.it (S.B.C.);
fabrizia.maccati@crea.gov.it (F.M.); elisabetta.lupotto@crea.gov.it (E.L.)
* Correspondence: laura.gennaro@crea.gov.it; Tel.: +39-065-149-4534

Academic Editor: Edgar Chambers IV
Received: 13 February 2017; Accepted: 30 June 2017; Published: 5 July 2017

Abstract: This work provides an updated picture of communication strategies developed to improve healthy dietary habits in schoolchildren, with a focus on the importance of milk consumption. The paper has investigated two main areas: the definition of the main orientations and key points of research approach relative to the communication methods, with special attention to multiple strategies and the identification of their peculiarities to increase daily milk consumption. The school environment is considered as a unique environment to help increase the adoption of a correct dietary habit and lifestyle; it increases physical activity by facilitating the flow of health-related information. In this regard, several studies have highlighted the importance and effectiveness of school-based interventions on a large-scale, also considering multiple contexts, early interventions as well as the involvement of teachers, students and families. The effective actions range from interventions on prices and the availability of desirable and undesirable foods to educational programmes that improve food knowledge and the choices of students and/or their parents. From the nutritional point of view, milk is an important component of a well-balanced diet—especially for children—because it contains essential nutrients. It is a substantial contributor to the daily energy intake; however, its consumption often declines with aging and becomes insufficient. Therefore, developing strategies to increase its consumption is an important objective to reach.

Keywords: children; healthy eating habits; milk; model and environments; school-based programmes

1. Introduction

Healthy eating during childhood and adolescence is very important both for the physical and cognitive development of individuals [1]. Eating behaviours change during the first years of life: children ask what, when, and how much to eat. They learn about food through direct experiences with food itself and by observing the eating behaviours modelled by others [2]. Food preferences derive from the interaction between genetic and environmental factors that leads to individual differences [3–5]. Food preferences represent the main predictor of food intake in children [6]. The benefits of food preferences is that the dislike for a food can be reduced or even reversed by a combination of several factors. The understanding of how these preferences are shaped through children's food experiences represents a key issue and is related to the several factors that influence children and adolescent dietary choices, i.e., early tasting exposure, availability and preference of particular foodtypes, portion size, parenting style, and modelling [7,8]. Eating behaviour is set up during early years of childhood and persists into adulthood: childhood represents a critical moment to establish healthy eating patterns [9–11].

A lot of studies analyse how parents and children influence each other's eating behaviour [2,12–15]. Both children's attitudes towards food and children's assessment of satiety could be affected by family [16,17].

Rhee et al. [18] describe three categories of parental influences: specific parent feeding practices, general parental behaviours and global parenting influences. The recent systematic review and meta-analysis of Yee et al. [19], by studying the influence of parental practices on child promotive and preventive food consumption behaviours, has highlighted how the number of parental behaviours are strong correlated to the dietary behaviour of children. However, the authors underlined three main understudied areas in parental behaviours that influence the patterns of food consumption among children: active guidance/education, psychosocial mediators and moderating influence of general parenting styles [19]. For example, some works highlight the increase of children's intake of fruit, vegetables and milk after observing adults or their peers consume these foods [20–23].

The positive modelling represents an indirect, yet effective, strategy to promote healthy eating patterns in children. Hebestreit et al. [24], throughout a European multi-centre study, pointed out how the parent-child communication approach is a key element of health education. In 2013/2014 the I. Family study cross-sectionally assessed the food intakes of families in eight European countries with to determine whether an association exists between children and parents' dietary patterns and whether the family food environment (the number of shared meals or soft drinks available during meals) affects this association: the availability of soft drinks and the negative parental role modelling resulted in important predictors of children's dietary patterns [25].

National and international organisms have elaborated guidelines to provide useful information about balanced diet and physical activity [26]. In the frame of the *"Guidelines for the Nutrition Education"* [27], for example, the Italian Ministry of Education has noted how an incisive line of action to improve children's dietary choices should involve the different environments in which the child spends most of his/her time: such as family, school, as well as healthcare and society settings. A large part of nutritional needs are met at school: starting from this assertion, school environment is the ideal place where policies can be implemented to encourage healthy eating habits in children [28–31]. This statment also underlies that teachers affect the eating behaviour of their students for a whole school year or many school years, acting as authoritative figures and representing a model through their own food habits [32–34].

For this reason, a lot of programmes have been developed at the school level and in other backgrounds [34–39]. In this perspective, the implementation of proper information campaigns is an important procedure to encourage the adoption of correct lifestyles, including physical activity. Their purpose is to give birth to integrated and multisectorial strategies and actions (projects, programmes, etc.) in school-based programs, involving different school and out-of-school figures (students, teachers, family, public bodies etc.). Their aim is to change the behaviour of the single or the group (educational approach) as well as the context that supports the incorrect behaviour (socio-ecological approach) [40].

2. Material and Methods

Our search strategy includes the use of search engines Scopus, Science Direct, and PubMed, where the following keywords were typed: communication strategies and children's dietary habits; dietary habits and childhood; teacher and children's dietary habits; family and children dietary habits; dietary habits and elementary school; fruit and communication strategies; vegetable and communication strategies; nutrition programmes and childhood.

For the focus on milk, additional keywords have been inserted: milk consumption and schoolchildren; flavoured milk consumption and schoolchildren; chocolate milk consumption and schoolchildren; communication strategies and children's milk consumption.

3. Attention to Multiple Strategies Is Needed to Improve Eating Habits in Children: Examples and Update of Communication Strategies

In this work, the main lines and key points of research approach—with attention to the multiple strategies—are described and extracted through updated and targeted examples, starting from the awareness that the effective actions for the promotion of children's healthy dietary habits involve a large number of school-based interventions: from interventions on prices and availability of desirable and undesirable foods to educational programmes to improve food knowledge and the choices of students and/or their parents [2,41–45].

From the meta-analysis of Dudley et al. [35] from the 49 eligible papers, it has emerged that the dominant strategies were: enhanced curriculum approach ($n = 29$), cross-curricular approaches ($n = 11$), parental involvement ($n = 10$), experiential learning approach ($n = 10$), contingent reinforcement approaches i.e., rewards or incentives given to students in response to desired behaviours ($n = 7$), literacy abstraction approaches i.e., literature read by/to children whereby a character promotes/exemplifies positive behaviours ($n = 3$), game-based approaches ($n = 2$), and web-based approaches ($n = 2$). It is also worth mentioning the recent review of Decosta et al. [46], where the following intervention areas were categorized: parent control, reward/instrumental feeding, social facilitation, cooking programmes, school gardens, sensory education and taste lessons, choice architecture and nudging, branding, food packaging, and spokes-characters, and offering a choice.

A first type of approach is the repetitive exposure to healthy foods [47]. As reported by Knai et al. [48] the increased exposure to healthy foods (i.e., fruit and vegetables) represents a key intervention, useful to push children towards healthier behaviours. Roe et al. [49], in a crossover design on 61 children (aged 3–5 years), showed that providing a variety of vegetables and fruit as snacks leads to an increased consumption of both food types in a childcare facility. However, motivation is another important factor for encouraging children to eat fruit and vegetables. In fact, some authors conclude that presenting a food as a reward or giving small prizes for tasting increases children's preference for that food, because they act as an incentive to encourage healthy eating in children [41,43,50]. On the contrary, pressing children to eat specific foods leads them to dislike those foods. Likewise, restricted access to some foods leads towards an overconsumption of those foods when children are free to choose them [51]. Again, Birch et al. [2] reported and underlined how giving rewards for food selection in preschool children could lead to avoid the same food when rewards are over. The recent work of Loewenstein et al. [52], instead, indicates that short-run incentives can produce changes in behaviour that persist after incentives are removed.

Concerning economic incentives, Jensen et al. [53] evaluated their effectiveness in eliciting sound nutritional behaviour in schools and highlighted how such incentives are effective for altering the consumption patterns in the school setting.

Another potential approach is the creation or implementation of food policies in the school environment that would limit or deny access to undesirable foods i.e., snacks and soft drinks. As reported by Cullen et al. [54], the availability of obesogenic foods in schools is associated with less healthy food preferences and food choices. Generally, the availability and accessibility of food represent the key determinants in food choices [55].

In a 2015 research, Losasso et al. [56] carried out a case study in Northeast Italy: several nutrition policies were developed in public schools. The purpose of the experiment was to compare the consumption of beverages and snacks in two different environments, school and extra-school contexts: the results highlighted the protective role of educational institutions in the promotion of healthier dietary patterns.

A dual strategy on the dietary behaviour of elementary pupils involving both a cafeteria environment intervention and a classroom nutritional program was reported by Song et al. [57] as a successful example of a multiple school-based nutrition program. However, in this work, the authors also have highlighted how behavioural economic approaches produce enhanced outcomes when paired with food education.

In this context, gardening can be considered as another advantageous strategy to encourage healthy dietary behaviours among children. It is worth mentioning the review of Berezowitz et al. [58] that summarizes studies on how school gardens could enhance the academic performance and the dietary habits in children: it offers evidence that garden-based learning does not negatively affect academic performance or fruit and vegetable consumption. On the contrary, it may favourably influence both. However considering the small set of studies, the authors also underline how other experimental designs and outcome measures are necessary [58]. The systematic review of Savoie-Roskos et al. [59] showed that although the evidence was mixed and fraught with limitations, most studies suggested how gardening-based interventions in school, community, or afterschool settings can have a small, but positive, influence on children's fruit/vegetables intake.

Another example of effective initiatives in schools is changing the length of time available for lunch. Cohen et al. [60] studied the association between the amount of time students (aged 8.4–15.6 years) have to eat their lunch and school food selection and consumption: having little time to eat was associated with a significant decrease in the consumption of entrées, milk, and vegetables.

This study suggests that school policies should encourage at least 25-min lunches in order to reduce food waste and improve dietary intake. Generally, schools should consider both collaborating with chefs and using a certain choice plan to increase fruit and vegetable selection, as reported in another study of Cohen et al. [61]; this research also stressed the importance not to abandon healthier options even if they are initially met with resistance [61]. The study of Just et al. [62] showed a pilot experiment of how chef-created dishes can increase school lunch participation and fruit and vegetable intake.

An effective approach of nutrition information aimed to increase healthy behaviour is the "image-based strategy", i.e., emolabeling, that uses simple expressions of emotions to convey a message on health (i.e., happy = healthy, sad = not healthy). Siegel et al. [63] demonstrated that using smiley emoticons in an elementary school cafeteria increased vegetable and Plain Fat Free Milk purchase (PFFM) by 29% and 141%, respectively. A combination between the use of both emoticons and small prizes was applied as a strategy in several studies. Barnes et al. [64], for example, showed how a two-level approach based on the use of emoticons followed by small prizes as incentives for better food choice led to significant and sustained improvement of healthy eating.

It is important to develop strategies that involve school staff, teachers, parents, and students in order to increase diet quality and physical activity [65–71]. Diet and physical inactivity are now among the leading causes of preventable death and disability, i.e., obesity [72]. Some examples of multiple and integrated strategies have been reported.

In a school-based intervention to promote healthy behaviours, Sacchetti et al. [73] highlighted the importance of planning integrated and multisectorial actions to encourage correct dietary habits and physical activity. The intervention involved 11 classes of children aged 8–11 for three years, from their third to their fifth class. The activities were planned, with the aim of modifying both the behaviour and the context that causes the incorrect behaviour, including training modules for teachers and sport instructors, educational activities in class, sport and games at school, cookery and sensory workshops, creation of didactic materials, and motor activities. Results have shown that dietary habits have improved after the intervention. The percentage of children who consumed an adequate mid-morning snack increased ($p < 0.0001$), while the percentage of children who consumed snacks and drinks after dinner decreased ($p < 0.01$). An increase, but not significant, of the percentage of children who ate five or more portions of fruits and vegetables daily was also reported. Furthermore, no significant modifications were observed in motor performances.

Another example of intervention strategy is shown by Moss et al. [74] and it is based on the quasi-experimental design that analyses the effects of the combination between the Coordinated Approach to Child Health (CATCH) [75] nutrition curriculum and the Farm to School program [76] to assess the nutrition knowledge of 3rd-grade students. CATCH, a coordinated school-based health program, was implemented by planning and introducing a curriculum on nutrition with educational activities—including physical activity—combining them with the Farm to School programme that encourages the consumption of locally grown foods to support farmers. Findings suggest that CATCH nutrition education and farm tours can positively affect nutrition knowledge and fruit and vegetable consumption behaviour among schoolchildren [74].

In this regard, it is interesting the work by Prelip et al. [77] that monitored the effects of a "hybrid" school-based nutrition programme on attitudes, beliefs, and behaviours related to fruit and vegetable consumption diffused throughout a large urban community. The hybrid intervention included a combination of district strategies, local school-defined strategies, and "home-made" strategies/activities created by teachers. The intervention resulted in a significant change in teacher influence on students' attitudes toward fruits and vegetables ($p < 0.05$) and students' attitudes towards vegetables ($p < 0.01$), even after adjusting for gender, grade, and race/ethnicity [77].

Considering that children today spend a large amount time on mobiles and social networks, and are exposed to various forms of interactive advertising [78], another strategy to encourage healthy habits is given by a combined approach on a school-based and media intervention. As an example, the work of Grassi et al. [79] showed the efficacy of this approach to increase fruit and vegetable intake in Italian children, while Blitstein et al. [80] underline the benefits of including a parent-focused social marketing campaign in nutrition education interventions. Additionally, it's worth mentioning examples of complementary online interventions. The recent work of Dumas et al. [81] showed an example of development of an evidence-informed blog to promote healthy eating among mothers using Intervention Mapping Protocol.

The recent study of Roccaldo et al. [82] showed how teachers' training program accompanying the "School Fruit Scheme" fruit distribution improves children's observance of the Mediterranean diet. This underlines the importance of the inclusion of teachers' training programmes in communication strategies to improve healthy eating habits in children.

At this regard, it is worth mentioning the study of Hall et al. [83] which highlights how health educators should collaborate with teachers in the design, implementation, and evaluation stages of curriculum development in order to better meet the needs of students and facilitate the delivery of high-quality nutrition education to them.

The work of Goldberg et al. [84] describes a school-based intervention, Great Taste, Less Waste (GTLW), a preliminary example of communication strategy that linked healthy eating to the environment to improve the quality of foods from home: GTLW was well received, but no significant changes were observed in the quality of food brought to school. Whether classrooms are an effective environment for change remains to be explored.

Additionally, the work of Kastorini et al. [85] is interesting, evaluating the effects, through a cohort study, of a food aid and promotion of healthy nutrition programme (the "DIATROFI" programme) on the diet quality of Greek students. At the end of the intervention, the consumption frequency of all the promoted foods, namely milk, fruits, vegetables, and whole grain products, increased among children and adolescents, boys and girls ($p \leq 0.002$). This study highlights the importance to address the school-based nutritional programmes in particular towards low socioeconomic status groups that tend to adopt unhealthier choices.

4. Some Additional Specificities Related to Strategies to Increase School Aged Children's Daily Milk Consumption

In the context of multiple and interdisciplinary communication approaches, the strategies to increase daily milk consumption among school children are placed. Identifying additional specificities for developing actions to increase the consumption of milk, as the main goal of this paper, is now discussed.

Milk is an important component of a well-balanced diet—especially for children—because it contains essential nutrients. It is a substantial contributor to the daily energy intake; however, its consumption often declines with aging and becomes insufficient [86]. The fundamental role of milk in school nutrition reflects its unique nutrient contributions to children's diets. The increase of consumption of milk at school could lead students closer to recommendations on nutrients provided by milk.

Encouraging people drink milk from a young age might represent a strategic action for increasing the intake of some essential nutrients and improving healthy habits. In order to encourage sales, milk is now produced with a set of characteristics that meet the needs of consumers and attract them (e.g., tasty flavours, new and captivating packaging, straws, etc.). Clearly, if milk is served with meals, the energy density and portion sizes of foods can influence children's energy intake. However Kling et al. [87] studied the effect of varying the energy density (1% fat or 3.25% fat) and the portion size of milk served with the lunch on the intake of milk among preschool children (3–5 years old). The results have shown that across all ages, food intake decreases when higher-energy density rather than lower-energy density milk is served, whereas meal energy intake (food plus milk) does not change significantly. This study highlighted how the effects of milk energy density on meal intake vary between children. In any case, serving milk in larger portions promotes the intake of this nutritious and dense beverage.

In the USA, in response to the increase of childhood overweight and obesity [88,89], milk with lower fat levels is requested by schools and at the same time U.S. companies are paying more attention to the reformulation of flavoured milk with significantly reduced added-sugar content. In this regard, the work of Li and Drake [90], understanding the sensory perception of both adults and children about flavoured milk can help food developers and manufacturers to achieve attractive attributes while reducing the sugar content to meet the needs of a healthy diet.

As for the availability of flavoured milk in school and relative communication strategies, the argumentation is controversial: several schools embrace the idea that any milk is better than no milk. In this case, flavoured milk represents an alternative for meeting the recommended intake of this food [91–93]. Other schools limit or ban the sale of chocolate milk. Several authors discuss if altering the availability of flavoured milk could influence other eating behaviours of students within meal compensation or after-school snacking patterns. Quann and Adams [94] reported that when flavoured milk was removed on 1 to all days of the week, there was a 26.0% reduction in milk sales and a 11.4% increase in the percentage of milk surplus, resulting in a 37.4% decrease in milk consumption. Waite et al. [95], by investigating how environmental intervention could affect elementary school students' food selection during lunchtime, concluded that requiring students to ask for an item rather than self-serve could help modify food choice. In detail, in one school, students were required to "Ask for Chocolate Milk", a strategy that resulted in an 18% increase in the selection of white milk compared to the choices made by control students. In another school, students were exposed to "Increased White Milk Quantity" (the availability of white milk was three times as much as that of chocolate), but the visual cue of a three-fold greater quantity of white milk compared to chocolate milk did not significantly alter selection patterns [95].

Yon et al. [96,97] reported how the flavoured milks reformulated to be either low-fat or with reduced sugars tested in a school programme, remained popular among students. Cohen et al. [98] reported how the effect of removing flavoured milk from schools during the 2012–2013 school year, among 1030 elementary and middle school children in Boston area, resulted in a reduction in the selection and consumption of total milk. Henry et al. [99] studied the impact of replacing traditional chocolate milk with the reduced-sugar option on milk consumption in elementary schools: children preferred chocolate milk over plain milk even when a reduced-sugar formula was offered; so, switching to reduced-sugar chocolate milk led to a decrease in the number of students choosing milk. Hanks et al. [100] showed how, in the context of students' lunch, eliminating chocolate milk from school cafeterias led to a reduction of calories and sugar consumption; but, at the same time, it led students to take less milk or drink less of the white milk they took; furthermore, the number of students participating in the National School Lunch Program decreased. Then, the authors added that there could also be some consequences on the way students compensated during lunch, or later in the day, for example selecting a dessert. The authors suggested that food service managers needed to carefully evaluate costs and benefits of removing chocolate milk and identify options to make white milk more convenient, attractive, and make it the default choice [100].

Other studies underlined that transforming "white milk" into the preferred choice, and making "flavoured milk" less convenient (without removing it), could lead to an immediate increase of white milk selection, by simultaneously reducing potential controversy [95,101,102].

Nowadays, also in some European countries (especially in Northern Europe) milk is directly taken from home. In some schools it can be sold by vending machines. Moreover, several programmes started to promote the distribution of milk in the schools for the mid-morning break or for lunch [103].

It is worth mentioning the work by List and Samek [104] that showed the positive impact of incentives adopted in the school lunchroom on the milk choice made by children from low-income households; the lunchroom is a "teachable moment" to encourage children in making healthy choices as it is defined by the same authors. The study of Sao et al. [105] underlined the importance of hands-on farming experience with dairy products in order to increase the consumption of milk and dairy products among children.

Hendrie et al. [106] have focused their attention on interventions targeting an increase in the consumption of dairy food or Ca intakes among children.

5. Conclusions

This work highlights the effectiveness of multiple and interdisciplinary communication strategies for improving healthy dietary habits among schoolchildren. It then concludes how the promotion of a well-balanced diet combined with physical activity should be the starting point of a modern strategy of communication. A fundamental step that needs to be integrated with other innovative actions, i.e., training courses for teachers, farm to school programs, and an enhanced curriculum approach.

The strategies to increase daily milk consumption among schoolchildren should follow this direction. Since the consumption of milk declines with aging and becomes insufficient, developing strategies to increase drinking of this beverage is an important goal to fix. As for additional specificities, the authors conclude that the promotion of milk with lower fat levels and the reformulation of flavoured milk with a significantly reduced added-sugar content should be considered.

Acknowledgments: The authors thank Annalisa Lista for the linguistic revision and the editing of this paper.

Conflicts of Interest: The authors declare no conflict of interest.

References

1. Jacka, F.N.; Kremer, P.J.; Berk, M.; de Silva-Sanigorski, A.M.; Moodie, M.; Leslie, E.R.; Pasco, J.A.; Swinburn, B.A. A Prospective Study of Diet Quality and Mental Health in Adolescents. *PLoS ONE* **2011**, *6*, e24805. [CrossRef] [PubMed]

2. Birch, L.; Savage, J.S.; Ventura, A. Influences on the development of children's eating: From infancy to adolescence. *Can. J. Diet. Pract. Res.* **2007**, *68*, s1–s56. [PubMed]

3. Breen, F.M.; Plomin, R.; Wardle, J. Heritability of food preferences in young children. *Physiol. Behav.* **2006**, *88*, 443–447. [CrossRef] [PubMed]

4. Llewellyn, C.H.; van Jaarsveld, C.H.; Boniface, D.; Carnell, S.; Wardle, J. Eating rate is a heritable phenotype related to weight in children. *Am. J. Clin. Nutr.* **2008**, *88*, 1560–1566. [CrossRef] [PubMed]

5. Gahagan, S. The Development of Eating Behavior—Biology and Context. *J. Dev. Behav. Pediatr.* **2012**, *33*, 261–271. [CrossRef] [PubMed]

6. Gibson, E.L.; Wardle, J.; Watts, C.J. Fruit and vegetable consumption, nutritional knowledge and beliefs in mothers and children. *Appetite* **1998**, *31*, 205–228. [CrossRef] [PubMed]

7. Patrick, H.; Nicklas, T.A. A review of family and social determinants of children's eating patterns and diet quality. *J. Am. Coll. Nutr.* **2005**, *24*, 83–92. [CrossRef] [PubMed]

8. Scaglioni, S.; Arrizza, C.; Vecchi, F.; Tedeschi, S. Determinants of children's eating behavior. *Am. J. Clin. Nutr.* **2011**, *94*, 2006S–2011S. [CrossRef] [PubMed]
9. Mikkilä, V.; Räsänen, L.; Raitakari, O.T.; Pietinen, P.; Viikari, J. Longitudinal changes in diet from childhood into adulthood with respect to risk of cardiovascular diseases: The Cardiovascular Risk in Young Finns Study. *Eur. J. Clin. Nutr.* **2004**, *58*, 1038–1045. [CrossRef] [PubMed]
10. Neumark-Sztainer, D.; Wall, M.; Larson, N.I.; Eisenberg, M.E.; Loth, K. Dieting and disordered eating behaviors from adolescence to young adulthood: Findings from a 10-year longitudinal study. *J. Am. Diet. Assoc.* **2011**, *111*, 1004–1011. [CrossRef] [PubMed]
11. Nicklaus, S.; Remy, E. Early origins of overeating: Tracking between early food habits and later eating patterns. *Curr. Obes. Rep.* **2013**, *2*, 179–184. [CrossRef]
12. Faith, M.S.; Scanlon, K.S.; Birch, L.L.; Francis, L.A.; Sherry, B. Parent-Child Feeding Strategies and Their Relationships to Child Eating and Weight Status. *Obes. Res.* **2004**, *12*, 1711–1722. [CrossRef] [PubMed]
13. Pearson, N.; Biddle, S.J.H.; Gorely, T. Family correlates of fruit and vegetable consumption in children and adolescents: A systematic review. *Public Health Nutr.* **2008**, *12*, 267–283. [CrossRef] [PubMed]
14. Anzman, S.L.; Rollins, B.Y.; Birch, L.L. Parental influence on children's early eating environments and obesity risk: Implications for prevention. *Int. J. Obes.* **2010**, *34*, 1116–1124. [CrossRef] [PubMed]
15. Briggs, L.; Lake, A.A. Exploring school and home food environments: Perceptions of 8–10-year-olds and their parents in Newcastle upon Tyne, UK. *Public Health Nutr.* **2011**, *14*, 2227–2235. [CrossRef] [PubMed]
16. Birch, L.L.; McPhee, L.; Shoba, B.C.; Steinberg, L.; Krehbiel, L. Clean up your plate: Effects of child feeding practices on the conditioning of meal size. *Learn. Motiv.* **1987**, *18*, 301–317. [CrossRef]
17. Nicklas, T.A.; Baranowski, T.; Baranowski, J.; Cullen, K.; Rittenberry, L.; Olvera, N. Family and child-care provider influences on preschool children's fruit, juice, and vegetable consumption. *Nutr. Rev.* **2001**, *59*, 224–235. [CrossRef] [PubMed]
18. Rhee, K. Childhood overweight and the relationship between parent behaviors, parenting style, and family functioning. *Ann. Am. Acad. Pol. Soc. Sci.* **2008**, *615*, 11–37. [CrossRef]
19. Yee, A.Z.; Lwin, M.O.; Ho, S.S. The influence of parental practices on child promotive and preventive food consumption behaviors: A systematic review and meta-analysis. *Int. J. Behav. Nutr. Phys. Act.* **2017**, *14*, 47. [CrossRef] [PubMed]
20. Raynor, H.A.; Van Walleghen, E.L.; Osterholt, K.M.; Hart, C.N.; Jelalian, E.; Wing, R.R.; Goldfield, G.S. The relationship between child and parent food hedonics and parent and child food group intake in children with overweight/obesity. *J. Am. Diet. Assoc.* **2011**, *111*, 425–430. [CrossRef] [PubMed]
21. Wang, Y.; Beydoun, M.A.; Li, J.; Liu, Y.; Moreno, L.A. Do children and their parents eat a similar diet? Resemblance in child and parental dietary intake: Systematic review and meta-analysis. *J. Epidemiol. Community Health* **2011**, *65*, 177–189. [CrossRef] [PubMed]
22. Draxten, M.; Fulkerson, J.A.; Friend, S.; Flattum, C.F.; Schow, R. Parental role modeling of fruits and vegetables at meals and snacks is associated with children's adequate consumption. *Appetite* **2014**, *78*, 1–7. [CrossRef] [PubMed]
23. Robinson, L.N.; Rollo, M.E.; Watson, J.; Burrows, T.L.; Collins, C.E. Relationships between dietary intakes of children and their parents: A cross-sectional, secondary analysis of families participating in the family diet quality study. *J. Hum. Nutr. Diet.* **2014**, *28*, 443–451. [CrossRef] [PubMed]
24. Hebestreit, A.; Keimer, K.M.; Hassel, H.; Nappo, A.; Eiben, G.; Fernández, J.M.; Kovacs, E.; Lasn, H.; Shiakou, M.; Ahrens, W. What do children understand? Communicating health behavior in a European multi-centre study. *J. Public Health* **2010**, *18*, 391–401. [CrossRef]
25. Hebestreit, A.; Intemann, T.; Siani, A.; Dehenauw, S.; Eiben, G.; Kourides, Y.A.; Kovacs, E.; Moreno, L.A.; Veidebaum, T.; Krogh, V.; et al. Dietary Patterns of European Children and Their Parents in Association with Family Food Environment: Results from the I. Family Study. *Nutrients* **2017**, *9*, 126. [CrossRef] [PubMed]
26. Istituto Nazionale di Ricerca per gli Alimenti e la Nutrizione (INRAN). *Linee Guida per Una Sana Alimentazione Italiana*; INRAN: Roma, Italy, 2003.
27. Ministero dell'Istruzione, dell'Università e della Ricerca (MIUR). *Linee Guida per L'educazione Alimentare, 2015. Direzione Generale per lo Studente, l'Integrazione e la Partecipazione. Aoodgsip. Registro Ufficiale(U)*; MIUR: Roma, Italy, 2015.

28. Larson, N.I.; Story, M.; Wall, M.; Neumark-Sztainer, D. Calcium and dairy intakes of adolescents are associated with their home environment, taste preferences, personal health beliefs, and meal patterns. *J. Am. Diet. Assoc.* **2006**, *106*, 1816–2431. [CrossRef] [PubMed]

29. Story, M.; Kaphingst, K.M.; French, S. The role of schools in obesity prevention. *Future Child* **2006**, *16*, 109–142. [CrossRef] [PubMed]

30. Lee, A. Health-promoting schools: Evidence for a holistic approach to promoting health and improving health literacy. *Appl. Health Econ. Health Policy* **2009**, *7*, 11–17. [CrossRef] [PubMed]

31. Fernández-Alvira, J.M.; Mouratidou, T.; Bammann, K.; Hebestreit, A.; Barba, G.; Sieri, S.; Reisch, L.; Eiben, G.; Hadjigeorgiou, C.; Kovacs, E.; et al. Parental education and frequency of food consumption in European children: The IDEFICS study. *Public Health Nutr.* **2012**, *12*, 1–12. [CrossRef] [PubMed]

32. Kubik, M.Y.; Lytle, L.A.; Hannan, P.J.; Story, M.; Perry, C.L. Food-related beliefs, eating behavior, and classroom food practices of middle school teachers. *J. Sch. Health* **2002**, *72*, 339–345. [CrossRef] [PubMed]

33. Prelip, M.; Erausquin, T.; Slusser, W.; Vecchiarelli, S.; Weightman, H.; Lange, L.; Neumann, C. The role of classrooms teachers in nutrition and physical education. *Calif. J. Health Promot.* **2006**, *4*, 116–127.

34. Perikkou, A.; Gavrieli, A.; Kougioufa, M.M.; Tzirkali, M.; Yannakoulia, M. A novel approach for increasing fruit consumption in children. *J. Acad. Nutr. Diet.* **2013**, *113*, 1188–1193. [CrossRef] [PubMed]

35. Dudley, D.A.; Cotton, W.G.; Peralta, L.R. Teaching approaches and strategies that promote healthy eating in primary school children: A systematic review and meta-analysis. *Int. J. Behav. Nutr. Phys. Act.* **2015**, *12*, 28. [CrossRef] [PubMed]

36. Lloyd, J.; Wyatt, K. The Healthy Lifestyles Programme (HeLP)—An Overview of and Recommendations Arising from the Conceptualisation and Development of an Innovative Approach to Promoting Healthy Lifestyles for Children and Their Families. *Int. J. Environ. Res. Public Health* **2015**, *12*, 1003–1019. [CrossRef] [PubMed]

37. Miller Lovell, C. 5-2-1-0 Activity and Nutrition Challenge for Elementary Students. *J. Sch. Nurs.* **2017**. [CrossRef]

38. Kessler, H.S. Simple interventions to improve healthy eating behaviors in the school cafeteria. *Nutr. Rev.* **2016**, *74*, 198–209. [CrossRef] [PubMed]

39. Foltz, J.L.; May, A.L.; Belay, B.; Nihiser, A.J.; Dooyema, C.A.; Blanck, H.M. Population-level intervention strategies and examples for obesity prevention in children. *Annu. Rev. Nutr.* **2012**, *32*, 391–415. [CrossRef] [PubMed]

40. Moore, L.; de Silva-Sanigorski, A.; Moore, S.N. A socio-ecological perspective on behavioural interventions to influence food choice in schools: Alternative, complementary or synergistic? *Public Health Nutr.* **2013**, *16*, 1000–1005. [CrossRef] [PubMed]

41. Horne, P.J.; Tapper, K.; Lowe, C.F.; Hardman, C.A.; Jackson, M.C.; Woolner, J. Increasing children's fruit and vegetable consumption: A peer-modelling and rewards-based intervention. *Eur. J. Clin. Nutr.* **2004**, *58*, 1649–1660. [CrossRef] [PubMed]

42. Kristjansdottir, A.G.; Johannsson, E.; Thorsdottir, I. Effects of a school-based intervention on adherence of 7–9-year-olds to food-based dietary guidelines and intake of nutrients. *Public Health Nutr.* **2010**, *13*, 1151–1161. [CrossRef] [PubMed]

43. Delgado-Noguera, M.; Tort, S.; Martínez-Zapata, M.J.; Bonfill, X. Primary school interventions to promote fruit and vegetable consumption: A systematic review and meta-analysis. *Prev. Med.* **2011**, *53*, 3–9. [CrossRef] [PubMed]

44. Boddy, L.M.; Knowles, Z.R.; Davies, I.G.; Warburton, G.L.; Mackintosh, K.A.; Houghton, L.; Fairclough, S.J. Using formative research to develop the healthy eating component of the CHANGE! School-based curriculum intervention. *BMC Public Health* **2012**, *12*, 710. [CrossRef] [PubMed]

45. Wang, D.; Stewart, D. The implementation and effectiveness of school-based nutrition promotion programmes using a health-promoting schools approach: A systematic review. *Public Health Nutr.* **2013**, *16*, 1082–1100. [CrossRef] [PubMed]

46. DeCosta, P.; Møller, P.; Frøst, M.B.; Olsen, A. Changing children's eating behaviour—A review of experimental research. *Appetite* **2017**, *113*, 327–357. [CrossRef] [PubMed]

47. Cooke, L. The importance of exposure for healthy eating in childhood: A review. *J. Hum. Nutr. Diet.* **2007**, *20*, 294–301. [CrossRef] [PubMed]

48. Knai, C.; Pomerleau, J.; Lock, K.; McKee, M. Getting children to eat more fruit and vegetables: A systematic review. *Prev. Med.* **2006**, *42*, 85–95. [CrossRef] [PubMed]
49. Roe, L.S.; Meengs, J.S.; Birch, L.L.; Rolls, B.J. Serving a variety of vegetables and fruit as a snack increased intake in preschool children. *Am. J. Clin. Nutr.* **2013**, *98*, 693–699. [CrossRef] [PubMed]
50. List, J.A.; Samek, A.S. The Behavioralist as Nutritionist: Leveraging Behavioral Economics to Improve Child Food Choice and Consumption. *J. Health Econ.* **2015**, *39*, 135–146. [CrossRef] [PubMed]
51. Savage, J.; Fisher, J.O.; Birch, L. Parental Influence on Eating Behavior: Conception to Adolescence. *J. Law Med. Ethics* **2007**, *35*, 22–34. [CrossRef] [PubMed]
52. Loewenstein, G.; Price, J.; Volpp, K. Habit formation in children: Evidence from incentives for healthy eating. *J. Health Econ* **2016**, *45*, 47–54. [CrossRef] [PubMed]
53. Jensen, J.D.; Hartmann, H.; de Mul, A.; Schuit, A.; Brug, J.; ENERGY Consortium. Economic incentives and nutritional behavior of children in the school setting: A systematic review. *Nutr. Rev.* **2011**, *69*, 660–674. [CrossRef] [PubMed]
54. Cullen, K.W.; Eagan, J.; Baranowski, T.; Owens, E.; de Moor, C. Effect of a la carte and snack bar foods at school on children's lunchtime intake of fruits and vegetables. *J. Am. Diet. Assoc.* **2000**, *100*, 1482–1486. [CrossRef]
55. Azétsop, J.; Joy, T.R. Access to nutritious food, socioeconomic individualism and public health ethics in the USA: A common good approach. *Philos. Ethics Hum. Med.* **2013**, *8*, 16. [CrossRef] [PubMed]
56. Losasso, C.; Cappa, V.; Neuhouser, M.L.; Giaccone, V.; Andrighetto, I.; Ricci, A. Students' consumption of beverages and snacks at school and away from school: A case study in the North East of Italy. *Front. Nutr.* **2015**, *2*, 30. [CrossRef] [PubMed]
57. Song, H.-J.; Grutzmacher, S.; Munger, A.L. Project ReFresh: Testing the Efficacy of a School-Based Classroom and Cafeteria Intervention in Elementary School Children. *J. Sch. Health* **2016**, *86*, 543–551. [CrossRef] [PubMed]
58. Berezowitz, C.K.; Bontrager Yoder, A.B.; Schoeller, D.A. School Gardens Enhance Academic Performance and Dietary Outcomes in Children. *J. Sch. Health* **2015**, *85*, 508–518. [CrossRef] [PubMed]
59. Savoie-Roskos, M.R.; Wengreen, H.; Durward, C. Increasing fruit and vegetable intake among children and youth through gardening-based interventions: A systematic review. *J. Acad. Nutr. Diet.* **2017**, *117*, 240–250. [CrossRef] [PubMed]
60. Cohen, J.F.W.; Jahn, J.L.; Richardson, S.; Cluggish, S.A.; Parker, E.; Rimm, E.B. Amount of time to eat lunch is associated with children's selection and consumption of school meal entrée, fruits, vegetables, and milk. *J. Acad. Nutr. Diet.* **2015**, *116*, 123–128. [CrossRef] [PubMed]
61. Cohen, J.F.W.; Richardson, S.A.; Cluggish, S.A.; Parker, E.; Catalano, P.J.; Rimm, E.B. Effects of Choice Architecture and Chef-Enhanced Meals on the Selection and Consumption of Healthier School Foods: A Randomized Clinical Trial. *JAMA Pediatr.* **2015**, *169*, 431–437. [CrossRef] [PubMed]
62. Just, D.R.; Wansink, B.; Hanks, A.S. Chefs move to schools. A pilot examination of how chef-created dishes can increase school lunch participation and fruit and vegetable intake. *Appetite* **2014**, *83*, 242–247. [CrossRef] [PubMed]
63. Siegel, R.M.; Anneken, A.; Duffy, C.; Simmons, K.; Hudgens, M.; Lockhart, M.K.; Shelly, J. Emoticon Use Increases Plain Milk and Vegetable Purchase in a School Cafeteria Without Adversely Affecting Total Milk Purchase. *Clin. Ther.* **2015**, *37*, 1938–1943. [CrossRef] [PubMed]
64. Barnes, A.S.; Hudgens, M.E.; Ellsworth, S.C.; Lockhart, M.K.; Shelley, J.; Siegel, R.M. Emoticons and Small Prizes to Improve Food Selection in an Elementary School Cafeteria: A 15 Month Experience. *J. Pediatr. Child Nutr.* **2016**, *2*, 100108.
65. Nemet, D.; Barkan, S.; Epstein, Y. Short- and long-term beneficial effects of a combined dietary-behavioral-physical activity intervention for the treatment of childhood obesity. *Pediatrics* **2005**, *115*, 443–449. [CrossRef] [PubMed]
66. Taylor, R.W.; McAuley, K.A.; Barbezat, W.; Strong, A.; Williams, S.M.; Mann, J.I. APPLE Project: 2-y findings of a community-based obesity prevention program in primary school age children. *Am. J. Clin. Nutr.* **2007**, *86*, 735–742. [PubMed]
67. Silveira, J.A.; Taddei, J.A.; Guerra, P.H.; Nobre, M.R.C. Effectiveness of school-based nutrition education interventions to prevent and reduce excessive weight gain in children and adolescents: A systematic review. *J. Pediatr.* **2011**, *87*, 382–392. [CrossRef]

68. Coleman, K.J.; Shordon, M.; Caparosa, S.L.; Pomichowski, M.E.; Dzewaltowski, D.A. The healthy options for nutrition environments in schools (Healthy ONES) group randomized trial: Using implementation models to change nutrition policy and environments in low income schools. *Int. J. Behav. Nutr. Phys. Act.* **2012**, *9*, 80. [CrossRef] [PubMed]

69. Katz, D.L.; Katz, C.S.; Treu, J.A.; Reynolds, J.; Njike, V.; Walker, J.; Smith, E.; Michael, J. Teaching healthful food choices to elementary school students and their parents: The Nutrition Detectives™ Program. *J. Sch. Health.* **2011**, *81*, 21–28. [CrossRef] [PubMed]

70. Martin, A.; Saunders, D.H.; Shenkin, S.D.; Sproule, J. Lifestyle intervention for improving school achievement in overweight or obese children and adolescents. *Cochrane Database Syst. Rev.* **2014**, *3*, CD009728.

71. Girelli, L.; Manganelli, S.; Alivernini, F.; Lucidi, F. A Self-determination theory based intervention to promote healthy eating and physical activity in school-aged children. *Cuade. Psicol. Deporte* **2016**, *16*, 13–20.

72. An, R. Diet quality and physical activity in relation to childhood obesity. *Int. J. Adolesc. Med. Health* **2017**, *1*, 29. [CrossRef] [PubMed]

73. Sacchetti, R.; Dallolio, L.; Musti, M.A.; Guberti, E.; Garulli, A.; Beltrami, P.; Castellazzi, F.; Centis, E.; Zenesini, C.; Coppini, C.; et al. Effects of a school based intervention to promote healthy habits in children 8–11 years old, living in the lowland area of Bologna Local Health Unit. *Ann. Ig.* **2015**, *27*, 432–446.

74. Moss, A.; Smith, S.; Null, D.; Long Roth, S.; Tragoudas, U. Farm to School and Nutrition Education: Positively Affecting Elementary School-Aged Children's Nutrition Knowledge and Consumption Behavior. *Child Obes.* **2013**, *9*, 51–56. [CrossRef] [PubMed]

75. Coordinated Approach to Child Health (CATCH). CATCH Is Working in 8500 Schools and Afterschool Programs. Available online: www.catchinfo.org/whats-catch/ (accessed on 19 November 2012).

76. National Farm to School Network. Nourishing Kids and Community. Available online: www.farmtoschool. org/aboutus.php/ (accessed on 16 November 2012).

77. Prelip, M.; Slusser, W.; Thai, C.L.; Kinsler, J.; Erausquin, J.T. Effects of a school-based nutrition program diffused throughout a large urban community on attitudes, beliefs, and behaviors related to fruit and vegetable consumption. *J. Sch. Health* **2011**, *81*, 520–529. [CrossRef] [PubMed]

78. Shin, W.; Huh, J.; Faber, R.J. Developmental antecedents to children's responses to online advertising. *Int. J. Advert.* **2012**, *31*, 719–740. [CrossRef]

79. Grassi, E.; Evans, A.; Ranjit, N.; Pria, S.D.; Messina, L. Using a mixed-methods approach to measure impact of a school-based nutrition and media education intervention study on fruit and vegetable intake of Italian children. *Public Health Nutr.* **2016**, *19*, 1952–1963. [CrossRef] [PubMed]

80. Blitstein, J.L.; Cates, S.C.; Hersey, J.; Montgomery, D.; Shelley, M.; Hradek, C.; Kosa, K.; Bell, L.; Long, V.; Williams, P.A.; et al. Adding a Social Marketing Campaign to a School-Based Nutrition Education Program Improves Children's Dietary Intake: A Quasi-Experimental Study. *J. Acad. Nutr. Diet.* **2016**, *116*, 1285–1294. [CrossRef] [PubMed]

81. Dumas, A.A.; Lemieux, S.; Lapointe, A.; Provencher, V.; Robitaille, J.; Desroches, S. Development of an Evidence-Informed Blog to Promote Healthy Eating Among Mothers: Use of the Intervention Mapping Protocol. *JMIR Res. Protoc.* **2017**, *6*, e92. [CrossRef] [PubMed]

82. Roccaldo, R.; Censi, L.; D'Addezio, L.; Berni Canani, S.; Gennaro, L. A teachers' training program accompanying the "School Fruit Scheme" fruit distribution improves children's adherence to the Mediterranean diet: An Italian trial. *Int. J. Food Sci. Nutr.* **2017**. [CrossRef] [PubMed]

83. Hall, E.; Chai, W.; Albrecht, J.A. A Qualitative Phenomenological Exploration of Teachers' Experience with Nutrition Education. *Am. J. Health Educ.* **2016**, *47*, 136–148. [CrossRef] [PubMed]

84. Goldberg, J.P.; Folta, S.C.; Eliasziw, M.; Koch-Weser, S.; Economos, C.D.; Hubbard, K.L.; Tanskey, L.A.; Wright, C.M.; Must, A. Great Taste, Less Waste: A cluster-randomized trial using a communications campaign to improve the quality of foods brought from home to school by elementary school children. *Prev. Med.* **2015**, *74*, 103–110. [CrossRef] [PubMed]

85. Kastorini, C.M.; Lykou, A.; Yannakoulia, M.; Petralias, A.; Riza, E.; Linos, A. on behalf of the DIATROFI Program Research Team. The influence of a school-based intervention programme regarding adherence to a healthy diet in children and adolescents from disadvantaged areas in Greece: The DIATROFI study. *J. Epidemiol. Community Health* **2016**, 1–7. [CrossRef]

86. Pereira, P.C. Milk nutritional composition and its role in human health. *Nutrition* **2014**, *30*, 619–627. [CrossRef] [PubMed]

87. Kling, S.M.R.; Roe, L.S.; Sanchez, C.E.; Rolls, B.J. Does milk matter: Is children's intake affected by the type or amount of milk served at a meal? *Appetite* **2016**, *105*, 509–518. [CrossRef] [PubMed]

88. Institute of Medicine (IOM). *Nutrition Standards for Foods in Schools: Leading the Way Toward Healthier Youth*; The National Academies Press: Washington, DC, USA, 2007.

89. Institute of Medicine (IOM). *School Meals: Building Blocks for Healthy Children*; The National Academies Press: Washington, DC, USA, 2010.

90. Li, X.E.; Drake, M. Sensory perception, nutritional role, and challenges of flavored milk for children and adults. *J. Food Sci.* **2015**, *80*, R665–R670. [CrossRef] [PubMed]

91. Murphy, M.M.; Douglass, J.S.; Johnson, R.K.; Spence, L.A. Drinking flavored or plain milk is positively associated with nutrient intake and is not associated with adverse effects on weight status in US children and adolescents. *J. Am. Diet. Assoc.* **2008**, *108*, 631–639. [CrossRef] [PubMed]

92. Davis, M.M.; Spurlock, M.; Ramsey, K.; Smith, J.; Beamer, B.A.; Aromaa, S.; McGinnis, P.B. Milk Options Observation (MOO): A Mixed-Methods Study of Chocolate Milk Removal on Beverage Consumption and Student/Staff Behaviors in a Rural Elementary School. *J. Sch. Nurs.* **2017**. [CrossRef]

93. Fayet-Moore, F. Effect of flavored milk vs. plain milk on total milk intake and nutrient provision in children. *Nutr. Rev.* **2016**, *74*, 1–17. [CrossRef] [PubMed]

94. Quann, E.E.; Adams, D. Impact on milk consumption and nutrient intakes from eliminating flavored milk in elementary schools. *Nutr. Today* **2013**, *48*, 127–134. [CrossRef]

95. Waite, A.; Goto, K.; Chan, K.; Giovanni, M.; Wolff, C. Do environmental interventions impact elementary school students' lunchtime milk selection? *Appl. Econ. Perspect. Policy* **2013**, *35*, 360–376.

96. Yon, B.A.; Johnson, R.K.; Stickle, T.R. School children's consumption of lower-calorie flavored milk: A plate waste study. *J. Acad. Nutr. Diet.* **2012**, *112*, 132–136. [CrossRef] [PubMed]

97. Yon, B.A.; Johnson, R.K. Elementary and middle school children's acceptance of lower calorie flavored milk as measured by milk shipment and participation in the National School Lunch Program. *J. Sch. Health* **2014**, *84*, 205–211. [CrossRef] [PubMed]

98. Cohen, J.F.W.; Richardson, S.; Parker, E.; Catalano, P.J.; Rimm, E.B. Impact of the New U.S. Department of Agriculture School Meal Standards on Food Selection, Consumption, and Waste. *Am. J. Prev. Med.* **2014**, *46*, 388–394. [CrossRef] [PubMed]

99. Henry, C.; Whiting, S.J.; Finch, S.L.; Zello, G.A.; Vatanparast, H. Impact of replacing regular chocolate milk with the reduced-sugar option on milk consumption in elementary schools in Saskatoon, Canada. *Appl. Phys. Nutr. Metab.* **2016**, *41*, 511–515. [CrossRef] [PubMed]

100. Hanks, A.S.; Just, D.R.; Wansink, B. Chocolate milk consequences: A pilot study evaluating the consequences of banning chocolate milk in school cafeterias. *PLoS ONE* **2014**, *9*, e91022. [CrossRef] [PubMed]

101. Patterson, J.; Saidel, M. The removal of chocolate milk in schools results in a reduction in total milk purchases in all grades, K-12. *J. Am. Diet. Assoc.* **2009**, *109*, A97. [CrossRef]

102. Just, D.; Price, J. Default options, incentives and food choices: Evidence from elementary-school students. *Public Health Nutr.* **2013**, *16*, 2281–2288. [CrossRef] [PubMed]

103. European Commission. European School Milk Scheme. Available online: http://ec.europa.eu/agriculture/milk/school-milk-scheme_en (accessed on 4 July 2017).

104. List, J.A.; Samek, A. A Field Experiment on the Impact of Incentives on Milk Choice in the Lunchroom. *Public Financ. Rev.* **2017**, *45*, 44–67. [CrossRef]

105. Seo, T.; Kaneko, M.; Kashiwamura, F. Changes in intake of milk and dairy products among elementary schoolchildren following experiential studies of dairy farming. *Anim. Sci. J.* **2013**, *84*, 178–184. [CrossRef] [PubMed]

106. Hendrie, G.A.; Brindal, E.; Baird, D.; Gardner, C. Improving children's dairy food and calcium intake: Can intervention work? A systematic review of the literature. *Public Health Nutr.* **2013**, *16*, 365–376. [CrossRef] [PubMed]

beverages

MDPI

Article

Small Prizes Increased Plain Milk and Vegetable Selection by Elementary School Children without Adversely Affecting Total Milk Purchase

Megan Emerson [1], Michelle Hudgens [2], Allison Barnes [2], Elizabeth Hiller [3], Debora Robison [3], Roger Kipp [3], Ursula Bradshaw [4] and Robert Siegel [2,*]

[1]   College of Medicine and Life Sciences, University of Toledo, Toledo, OH 43614, USA;
      Megan.Emerson@rockets.utoledo.edu
[2]   Center for Better Health and Nutrition, Cincinnati Children's Hospital Medical Center, Cincinnati,
      OH 45229, USA; Michelle.Hudgens@cchmc.org (M.H.); Allison.Barnes@cchmc.org (A.B.)
[3]   Norwood City School District, Norwood, OH 45212, USA; hiller.e@norwoodschools.org (E.H.);
      Robison.D@norwoodschools.org (D.R.); kipp.r@norwoodschools.org (R.K.)
[4]   James M. Anderson Center for Health Systems Excellence, Cincinnati Children's Hospital Medical Center,
      Cincinnati, OH 45229, USA; ursula.bradshaw@cchmc.org
*   Correspondence: bob.siegel@cchmc.org; Tel.: +11-513-608-3243

Academic Editor: Alessandra Durazzo
Received: 3 October 2016; Accepted: 13 February 2017; Published: 17 February 2017

Abstract: (1) Background: Pediatric obesity continues to be a major public health issue. Poor food selection in the school cafeteria is a risk factor. Chocolate or strawberry flavored milk is favored by the majority of elementary school students. Previous health promotion efforts have led to increased selection of plain milk, but may compromise total milk purchased. In our study, we examined the effectiveness of small prizes as incentives to improve healthy food and beverage selection by elementary school students; (2) Methods: In a small Midwestern school district, small prizes were given to elementary school students who selected a "Power Plate" (PP), the healthful combination of a plain milk, a fruit, a vegetable and an entrée with whole grain over two academic school years; (3) Results: PP selection increased from 0.05 per student to 0.19, a 271% increase ($p < 0.001$). All healthful foods had increased selection with plain milk having the greatest increase, 0.098 per student to 0.255, a 159% increase ($p < 0.001$); (4) Total milk purchased increased modestly from 0.916 to 0.956 per student ($p = 0.000331$). Conclusion: Giving small prizes as a reward for healthful food selection substantially improves healthful food selection and the effect is sustainable over two academic years.

Keywords: milk; obesity; small prizes

1. Introduction

Pediatric obesity continues to be a major worldwide health problem [1,2]. Poor food selection such as choosing flavored milk or not selecting fruits and vegetables in the cafeteria by elementary school students is a risk factor for obesity [3]. Milk consumption in children is associated with improved Body Mass Index (BMI) status and positive intake of essential nutrients [4]. Flavored milk, which has added sugar, is offered in many US schools as part of the US Department of Agriculture free and reduced lunch program. Plain milk is also offered but flavored milk is preferred by the majority of students [5].

Initiatives to promote better food selection such as product placement, attractive display and featured naming (foods are given playful names) and convenience lines (those selecting healthful choices go to a shorter check-out line) typically improve healthy food selection by 20% to 100% [6,7]. Privatera et al. showed that emoticons can be used to influence school aged children's food choices [8].

Incentives to improve adult lifestyle behavior have been used successfully for several health goals including weight loss, smoking cessation and global wellness initiatives [9–11]. McDonald's corporation successfully introduced the "Happy Meal" concept in 1978 to increase sales to children [12]. Hobin et al. used small prizes to increase healthful packaged meal selection by 100% [13]. Just et al. demonstrated that school aged children would increase fruits and vegetables selected at lunch by 80% if given an incentive [14]. This type of intervention was further explored by List and Samek who demonstrated that when small prizes were offered to school aged children in separate interventions, plain milk selection increased from 16% to 40% and a more healthful snack from 17% to 75% of students [15,16].

Our research group had previously piloted the use of emoticons alone and then emoticons with small prizes to increase selection of healthful food selection in a Cincinnati inner city elementary school [17,18]. With the placement of "green smiley-faced" emoticons alone, plain fat-free milk and vegetables increased by 141% and 29% respectively, but there was no improvement in fruit or entrée selection [17]. To further increase healthful selection, we then added the Power Plate Program (PPP) to the emoticons [18]. The "Power Plate" (PP) was defined as the most healthful combination of foods available within the school lunch program: a plain milk, a vegetable, a fruit and an entrée with whole grain (sold together). With the "Power Plate Program" (PPP), emoticons were placed near the individual PP food items. While children still selected food items separately, they received a small prize for selecting the PP combination of a fruit, a vegetable, an entrée with whole grain and plain milk on PP days. Prizes consisted of a sticker, temporary tattoo, or a small toy valued at 40 cents or less. During the initial 10 weeks of the PPP, selection of plain milk increased by over 500%, vegetable selection by 71% and fruit selection by 20%. Frequency of "PP Days" (twice a week versus everyday) and quality of the prizes were tested and had little effect. At 15 months follow up, plain milk selection was still 253% above baseline when stickers or tattoos were given twice a week [19].

Based on our initial pilots, we determined the addition of the small prizes to the emoticons increased healthful food selection. Additionally, we tested whether the PPP altered the amount of waste by students and found there was no significant difference in waste of any of the PP food items [20]. Finally, we piloted the PPP in the three elementary schools of the Norwood City School District (NCSD) which has a diverse and economically broad student body. During this brief three week pilot, PP selection increased by over 500% [21]. Thus the work of other investigators and our group show that emoticon labelling and small prizes as incentives can be used to improved healthful food purchase by elementary school age children. With our current study we describe using emoticons and small prizes on larger scale over two academic years in an ethnically diverse school system.

2. Materials and Methods

All three elementary schools in the Norwood City School District were recruited for the extended PPP. The demographics of the schools are described in Table 1.

Table 1. Demographics of schools participating in intervention.

School	Enrollment	%Black	%White	%Hispanic	%Girls	%Boys	%Low Income *
Norwood View	409	12.7	68.7	13.9	48	52	70.4
Sharpsburg Elementary	260	12.7	71.2	8.8	44	56	72.7
Williams Avenue	291	12.4	75.3	6.5	48	52	75.9
All Schools Combined	960	12.6	71.3	10.2	47	53	72.7

Source: [22]. * Less than 130% of poverty level.

Children in grades K through 6 participated in the PPP at Norwood View and Williams Avenue Elementary Schools. Only grades 3 through 6 of the Sharpsburg Elementary School participated as grades K to 2 have a separate lunch room that does not allow for children to select their own food items. For those participating in the NCSD Lunch Program, children must select an entrée with whole grain (these are served together) and a vegetable. The students have the option to select up to two fruits, a plain milk or chocolate milk and an additional vegetable if available. Thus on most days, students may select up to two vegetables (with a minimum of one) and two fruits per day. The PPP was rolled out in the three schools as follows: Norwood View Elementary on 3 November 2014, Sharpsburg Elementary on 5 January 2015, and Williams Ave. Elementary School 13 April 2015. The cafeteria cash registers at the schools were equipped with a button to record when a student selected the PP of fruit, vegetable, plain milk and entrée with whole grain. PP data collection was started four weeks before the PPP began at each of the schools. During the first week of the program, "green-smiley faced" emoticons were placed by the PP food items and beverages. Bracelets were given out on the first day if a PP was selected. Figure 1 demonstrates the "green smiley-faced" emoticon placed near the PFFM (a) and in (b) bracelets that were given out a small prize during the "kick-off" week with a PP that was selected by one of the students.

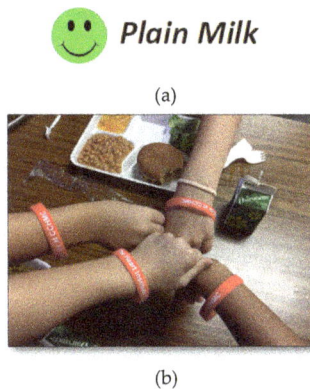

(a)

(b)

Figure 1. Emoticon example and students wearing "kick off bracelets" by a Power Plate (PP): (**a**) Plain Milk Emoticon; (**b**) Students wearing bracelets by PP.

Thereafter, on Tuesdays and Thursdays, stickers or temporary tattoos were given to students who selected the PP combination for the remainder of the academic school year by cafeteria staff and volunteers. During the PPP, students still had to choose food and beverage items independently, but were rewarded on PP days if they chose the PP (a fruit, a vegetable, an entrée with whole grain and a plain milk). The PPP was resumed for all three schools in September 2015 during the second academic year of the program and continued through the end of May2016. Thus the intervention lasted for 56 weeks with a 3-week gap at the beginning of the 2nd academic year during weeks 23 to 26. Cafeteria cash register receipt data was collected for one month prior to the PPP intervention and then throughout the entire intervention period. Data were obtained for all food items except for entrée as all students who purchase a lunch receive automatically an entrée with whole grain. Purchase data obtained from cash register receipts were supplied by the Food Services Department of Norwood City School District (NCSD) Statistical analysis by Z-testing (OpenEpi, Version 3, http://www.openepi.com) was used to compare the differences in the rate of food item selected per student per day between the baseline period (no intervention) to the intervention period when emoticons plus PP prizes were used. The NCSD was involved in the design, implementation and analysis of our intervention. This project was reviewed by the Cincinnati Children's Institutional Review Board, determined not to be a Human Subjects Research project and thus exempt.

3. Results

Observations were made on 158,596 school lunches purchased between October 2, 2014 and May 25, 2016 at the three elementary schools. Table 2 lists the PP and individual food item purchases comparing baseline (one month before the PPP was initiated) and throughout the intervention period to the end of the 2nd academic year. The rate is the number of a particular food item or beverage selected per student per lunch time. There were favorable increases reported for the PP (271%), plain milk (159%), fruits (18%), and vegetables (9%). Chocolate milk purchases decreased by 14% and total milk (plain + chocolate milk) had a more modest, but significant increase of 4%. The increase in vegetables selected, however, was somewhat surprising as students who participate in the school lunch program are automatically served a vegetable and entrée with whole grain in the NCSD schools at baseline. Selection of fruit and selection of the type of milk is at the student's discretion. Since students are typically offered two fruits and two vegetables with each lunch, it was possible for students to have greater than one vegetable or fruit serving with each meal. There was a dip in PP selection at weeks 20 to 25 which corresponds to the end of the first academic year when there were no cafeteria volunteers to help with the program and the beginning of the second academic before the PPP was resumed.

Table 2. Summary of Food Items selected by Norwood elementary students comparing pre-incentive rate and after initiating incentives. Z-tests were performed to determine significance between purchase rate before and during the intervention.

Food Item	Rate Selected Per Student Pre-Incentives n = 13,506 lunches	95% Confidence Interval	Rate selected Per Student With Incentives n = 145,090 lunches	95% Confidence Interval	p Value Comparing Pre to with Incentives
Power Plate *	0.0522	0.05-0.06	0.194	0.19-0.20	<0.0000001
Plain milk *	0.0984	0.09-0.11	0.255	0.25-0.26	<0.0000001
Chocolate milk *	0.818	0.80-0.84	0.700	0.70-0.71	<0.0000001
Total (Chocolate + Plain) milk *	0.916	0.90-0.94	0.956	0.95-0.96	0.000331
Vegetables *	1.299	1.27-1.32	1.416	1.41-1.42	<0.0000001
Fruits *	0.667	0.65-0.69	0.790	0.79-0.80	<0.0000001

* Experienced desirable, significant change comparing pre-intervention to intervention ($p < 0.05$).

Mean PP section increased from 0.0522 PP per student before incentives to 0.194 per student after implementing the incentives twice a week for PP selection. The graph is displayed with the baseline data period and PP initiation in phase (note: the initiation dates were different for the schools). Thus, week number is displayed rather than date. While the x-axis of the graph is numbered by week, each individual point represents a single day. The "zig-zag" pattern of the graph reflects increased PP sales on Tuesdays and Thursdays when prizes were given and decreased PP sales on non-prize days. The red "Mean" line shows the means for the entire pre-intervention (baseline) and the entire intervention period.

Figure 2 displays percent of students selecting the PP versus time at the three school individually during the 2014–15 academic year. Even though the dates of introduction are staggered, the effect is similar at each school suggesting that time of year is not a factor in PP selection. Events are annotated such as baseline period, when the PP was initiated and events that may have influenced the effectiveness of the program such as cafeteria staff changes at the Sharpsburg School or when the school dietary interns who helped run program left for summer break.

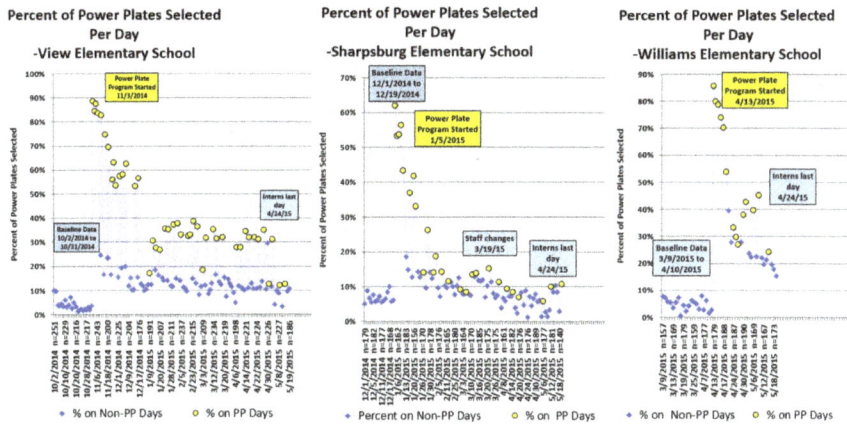

Figure 2. Shows daily PP purchases at the individual school during the first academic year. Each point represents a single day. A "zig-zag" effect is noted with increased PP selection on prize days. The Sharpsburg School which had a change in cafeteria staff experienced the sharpest decline in PP selection towards the end of the school year.

4. Discussion

This study shows that an incentive program to encourage school children to choose healthier food and beverage items in the lunch cafeteria was successful and the effect maintained on days prizes were given across a substantial time period and across schools with a diverse student body. The intervention resulted in a 5-fold increase in the selection of healthier items and this effect continued across two academic years. The intervention, which involved twice weekly provision of stickers or tattoos as 'prizes' for choosing healthier items (as indicated with a green happy face emoticon), compares favorably to other types of interventions to improve food selection in school cafeterias which typically see increases in the range of 20% to 100% [6,7,14]. Specifically, plain milk selection experienced the largest increase observed at 150% over baseline sales of plain milk. Given that about 60% of the 960 students participate in the school lunch program, this translates into an increase of about 100 cartons of plain milk per day for all the schools combined. Fruits had an 18% increase during the PPP compared to baseline. The smallest individual effect was with vegetables which increased by 9 percent over baseline purchase of vegetables. This increase is still remarkable as students are automatically served a vegetable by the school cafeteria staff.

Milk is recognized as an important source of calcium and other nutrients in children [23]. The majority of elementary school children select chocolate milk over plain milk when participating in the USDA school lunch program [24]. While there is some uncertainty on how chocolate milk selection over plain milk selection affects health status, chocolate milk contains about twice the sugar as plain milk. Attempts at eliminating chocolate milk and only offering plain milk in school cafeterias have led to increased plain milk purchase, but with a 10% to 26% overall milk purchase decrease and a decrease in consumption [25,26]. In our intervention, we successfully increased plain milk and even had a modest increase in overall milk purchase.

There are several limitations to our study. There are concerns as described previously by Birch et al. that giving external rewards for food selection may lead to avoiding a particular food when the rewards are stopped [27,28]. We did see PP purchases drop on days that rewards were not given with PP sales remaining marginally higher than baseline PP sales. Even with our extended intervention children reverted to close to their baseline choices on days without the incentives suggesting that the intervention is useful for changing foods purchased/chosen but not sufficient for changing preferences. Follow-up data post-intervention would show whether this marginal increase is sustained. Of concern, when the PPP was stopped at the end of the first academic year, there was a drop in PP selection. Also, this study did not evaluate the impact of the intervention on the overall diet of the children. Further, we only had purchase data and did not measure actual food / beverage consumption. We cannot comment on how consumption was affected or how individual purchases varied during the study. However, consumption data using the PPP in a previous inner city elementary school pilot showed that waste was unaffected by the program [20].

There are practical issues for implementation and sustainability of the PPP. The program was administered by existing cafeteria staff and volunteers. While this makes the PPP inexpensive to operate, school cafeterias are a busy environment and staff and volunteers may be distracted by other duties. This program was successfully implemented using the research format (i.e., with existing cafeteria staff and volunteers) over two academic years but that the data from one of the schools showed that cafeteria staff changes impacted negatively on the effectiveness of the PPP. Therefore scaling up of this intervention to more schools or even state-wide might need support mechanisms when cafeteria staff changes occur. The strengths of our study are that we have a very large observation of lunches purchased directly through cafeteria receipts over a two academic years. Given that the PPP was implemented with existing resources, the program could be cost-effective and sustainable when scaled up to a state-wide level.

5. Conclusions

We conclude that the PPP is an effective intervention in increasing healthful food selection in elementary school children. The greatest effect was in plain milk replacing flavored milk purchased.

Acknowledgments: We would like to thank the staff and students of the Norwood City School District for their help with this project. The project described was supported by the National Center for Advancing Translational Sciences of the National Institutes of Health, under Award Number 1UL1TR001425-01. The content is solely the responsibility of the authors and does not necessarily represent the official views of the NIH.

Author Contributions: All authors listed contributed substantially to this project and manuscript. R Siegel, M. Hudgens, D. Robison and R Kipp conceived and designed the experiments; A. Barnes, E. Hiller and M. Hudgens performed the experiments; R. Siegel, M. Emmerson, Ursula Bradshaw and M. Hudgens analyzed the data; M. Emmerson and U. Bradshaw contributed analysis tools; M. Emmerson, M. Hudgens, A. Barnes, R. Kipp, E. Hiller and R. Siegel wrote and reviewed the paper.

Conflicts of Interest: The authors declare no conflict of interest. The founding sponsors had no role in the design of the study; in the collection, analyses, or interpretation of data; in the writing of the manuscript, and in the decision to publish the results.

References

1. Ng, M.; Fleming, T.; Robinson, M.; Thomson, B.; Graetz, N.; Margono, C.; Mullany, E.C.; Biryukov, S.; Abbafati, C.; Abera, S.F.; et al. Global, regional, and national prevalence of overweight and obesity in children and adults during 1980–2013: A systematic analysis for the Global Burden of Disease Study 2013. *Lancet* **2014**, *384*, 766–781. [CrossRef]
2. Ogden, C.L.; Carroll, M.D.; Kit, B.K.; Flegal, K.M. Prevalence of childhood and adult obesity in the United States, 2011–2012. *JAMA* **2014**, *311*, 806–814. [CrossRef] [PubMed]
3. Finkelstein, D.M.; Hill, E.L.; Whitaker, R.C. School food environments and policies in US public schools. *Pediatrics* **2008**, *122*, e251–e259. [CrossRef] [PubMed]
4. Zheng, M.; Rangan, A.; Allman-Farinelli, M.; Rohde, J.F.; Olsen, N.J.; Heitmann, B.L. Replacing sugary drinks with milk is inversely associated with weight gain among young obesity-predisposed children. *Br. J. Nutr.* **2015**, *114*, 1448–1455. [CrossRef] [PubMed]
5. Hutchins, E. Flavored Milk and the National School Lunch Program. Available online: http://uknowledge. uky.edu/cph_etds/23 (accessed on 11 October 2016).
6. French, S.A.; Stables, G. Environmental Interventions to Promote Vegetable and Fruit Consumption Among Youth in School Settings. *Prev. Med.* **2003**, *37*, 593–610. [CrossRef] [PubMed]
7. Hanks, A.S.; Just, D.R.; Smith, L.E.; Wansink, B. Healthy convenience: Nudging students toward healthier choices in the lunchroom. *J. Pub. Health* **2012**, *34*, 370–376. [CrossRef] [PubMed]

8. Privitera, G.J.; Taylor, E.P.; Misenheimer, M.; Paque, R. The effectiveness of "emolabeling" to promote healthy food choices in children preschool through 5th grade. *Int. J. Child. Health Nutr.* **2014**, *3*, 48–54. [CrossRef]

9. Cawley, J.; Price, J.A. A case study of a workplace wellness program that offers financial incentives for weight loss. *J. Health Econ.* **2013**, *32*, 794–803. [CrossRef] [PubMed]

10. Volpp, K.G.; Troxel, A.B.; Pauly, M.V.; Glick, H.A.; Puig, A.; Asch, D.A.; Galvin, R.; Zhu, J.; Wan, F.; DeGuzman, J.; Corbett, E. A randomized, controlled trial of financial incentives for smoking cessation. *N. Engl. J. Med.* **2009**, *360*, 699–709. [CrossRef] [PubMed]

11. Baicker, K.; Cutler, D.; Song, Z. Workplace wellness programs can generate savings. *Health Aff. (Millwood)* **2010**, *29*, 304–311. [CrossRef] [PubMed]

12. Brownell, K.; Horgen, K. *Food Fight: The Inside Story of the Food Industry, America's Obesity Crisis, and What We Can Do About It*; Contemporary Books: Chicago, IL, USA, 2004.

13. Hobin, E.P.; Hammond, D.G.; Daniel, S.; Hanning, R.; Manske, S.R. The Happy Meal® effect: The impact of toy premiums on healthy eating among children in Ontario, Canada. *Can. J. Public Health* **2012**, *103*, e244–e248. [PubMed]

14. Just, D.R.; Price, J. Using incentives to encourage healthy eating in children. *J. Human Resour.* **2013**, *48*, 855–872. [CrossRef]

15. List, J.A.; Samek, A.S. The behavioralist as nutritionist: Leveraging behavioral economics to improve child food choice and consumption. *J. Health Econ.* **2015**, *39*, 135–146. [CrossRef] [PubMed]

16. List, J.A.; Samek, A. A field experiment on the impact of incentives on milk choice in the lunchroom. *Public Finan. Rev.* **2015**, *45*, 44–67. [CrossRef]

17. Siegel, R.M.; Anneken, A.; Duffy, C.; Simmons, K.; Hudgens, M.; Lockhart, M.K.; Shelly, J. Emoticon use increases plain milk and vegetable purchase in a school cafeteria without adversely affecting total milk purchase. *Clin. Ther.* **2015**, *37*, 1938–1943. [CrossRef] [PubMed]

18. Siegel, R.M.; Hudgens, H.; Annekin, A.; Simmons, K.; Shelly, J.; Bell, I.; Kotagal, U.R. A Two-Tiered School Cafeteria Intervention of Emoticons and Small Prizes Increased Healthful Food Selection. *IJFANS*. **2016**, 5. Available online: http://www.ijfans.com/ijfansadmin/upload/ijfans_5784ca415b87a.pdf (accessed on 12 October 2016).

19. Barnes, A.S.; Hudgens, M.E.; Ellsworth, S.C.; Lockhart, M.K.; Shelley, J.; Siegel, R.M. Emoticons and Small Prizes to Improve Food Selection in an Elementary School Cafeteria: A 15 Month Experience. *J. Pediatr. Child Nutr.* **2016**, *2*, 100108.

20. Hudgens, M.; Barnes, A.; Lockhart, M.K.; Ellsworth, S.; Beckford, M.; Siegel, R. Small Prizes Improve Food Selection in a School Cafeteria without Increasing Waste. *Clin. Pediatr.* **2016**. [CrossRef] [PubMed]

21. Siegel, R.; Lockhart, M.K.; Barnes, A.S.; Hiller, E.; Kipp, R.; Robison, D.L.; Ellsworth, S.C.; Hudgens, M.E. Small prizes increased healthful school lunch selection in a Midwestern school district. *Appl. Physiol. Nutr. Metab.* **2014**, *41*, 370–374. [CrossRef] [PubMed]

22. Norwood City School District. Available online: http://public-schools.startclass.com/d/d/Norwood-City-%28District%29 (accessed on 16 February 2017).

23. Ellery, J. The nutritional importance of milk and milk products in the national diet. *Int. J. Dairy Tech.* **1978**, *31*, 179–181. [CrossRef]

24. Fayet-Moore, F. Effect of flavored milk vs plain milk on total milk intake and nutrient provision in children. *Nutr. Rev.* **2016**, *74*, 1–7. [CrossRef] [PubMed]

25. Quann, E.E.; Adams, D. Impact on milk consumption and nutrient intakes from eliminating flavored milk in elementary schools. *Nutr. Today* **2013**, *48*, 127–134. [CrossRef]

26. Hanks, A.S.; Just, D.R.; Wansink, B. Chocolate milk consequences: A pilot study evaluating the consequences of banning chocolate milk in school cafeterias. *PLoS ONE* **2014**, *9*, e91022. [CrossRef] [PubMed]

27. Birch, L.L.; Birch, D.; Marlin, D.W.; Kramer, L. Effects of Instrumental Eating on Children's Food Preferences. *Appetite* **1982**, *3*, 125–134. [CrossRef]

28. Birch, L.L.; Marlin, D.W.; Rotter, J. Eating as the "means" activity in a contingency: Effects on young children's food preference. *Child Dev.* **1984**, *55*, 431–439. [CrossRef]

beverages

MDPI

Article

Reactions to a Low-Fat Milk Social Media Intervention in the US: The *Choose 1% Milk* Campaign

Robert John [1],*, Karla J. Finnell [1], Dave S. Kerby [1], Jade Owen [1] and Kendra Hansen [2]

[1] Health Sciences Center, University of Oklahoma, Oklahoma City, OK 73126-0901, USA;
karla-finnell@ouhsc.edu (K.J.F.); dave.s.kerby@gmail.com (D.S.K.); jade-owen@ouhsc.edu (J.O.)
[2] Two Rivers Public Health Department, Suite A, Kearney, NE 68847, USA; khansen@trphd.org
* Correspondence: robert-john@ouhsc.edu; Tel.: +1-405-271-2017 (ext. 46755)

Academic Editor: Alessandra Durazzo
Received: 31 May 2017; Accepted: 23 August 2017; Published: 25 September 2017

Abstract: (1) Background: Social media has increased in importance as a primary source of health communication but has received little academic attention. The purpose of this study was to conduct a content analysis of Facebook comments made in response to a five-week statewide social media intervention promoting use of 1% low-fat milk. Formative research identified health messages to promote, and 16 health messages consistent with the Dietary Guidelines for Americans were posted. During the intervention, 454 Facebook users posted 489 relevant comments; (2) Methods: The themes of user comments were identified using mixed-methods with qualitative identification of themes supplemented by cluster analysis; (3) Results: Six broad themes with 19 sub-themes are identified: (a) sugar, fat, and nutrients, (b) defiant, (c) watery milk, (d) personal preference, (e) evidence and logic, and (f) pure and natural; (4) The subject of milk is surprisingly controversial, a contested terrain in the mind of the consumer with a variety of competing perspectives that influence consumption. Public reactions to a social media nutrition education intervention are useful in understanding audience psychographics toward the desired behavior, require continual efforts to monitor and manage the social media campaign, but provide an opportunity to maximize the utility of real-time interactions with your audience.

Keywords: milk; social media; social marketing; Supplemental Nutrition Assistance Program (SNAP); cluster analysis; audience psychographics

---

1. Introduction

Saturated fat is an over-consumed nutrient in the American diet [1]. Since 1990, the Dietary Guidelines for Americans have been trying to get the US population to adopt 1% low-fat milk to reduce saturated fat consumption with limited success [2], and the recommendation to consume nonfat or 1% low-fat milk has become more explicit and more pervasive in each iteration of the Dietary Guidelines [3–7].

However, the evidence is that American consumers, by and large, have not heeded the recommendation. Figure 1 displays the 40-year trend in fluid milk sales in the US [8]. From 1975 to the early 1990s, whole milk sales plummeted as 2% milk sales increased while sales of other types of milk increased slightly during the late 1980s. In 1993, 2% milk sales exceeded whole milk sales for the first time and for the next decade whole and 2% milk sales were nearly identical. Since 2005, 2% milk sales have consistently surpassed whole milk sales. As of 2015, 1% low-fat and nonfat milk represents less than a third of all milk sales (30.0%), a figure that has not changed markedly during the last decade.

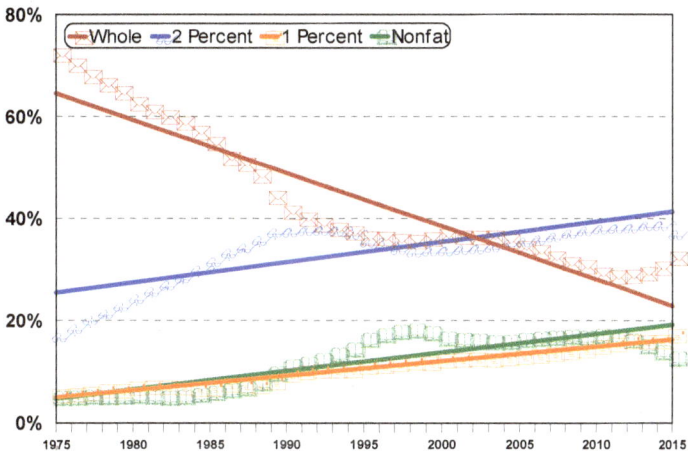

Figure 1. Actual and Trends in Fluid Milk Sales 1975–2015. Source. US Department of Agriculture; Adapted from: Fluid milk sales by product (Annual), 2016 [8].

Unfortunately, the shift toward 2% milk consumption does not achieve the aim of the Dietary Guidelines for Americans to reduce consumption of saturated fat. Replacing whole and 2% milk with nonfat and 1% milk significantly reduces the amount of saturated fat and calories consumed at a population level [9]. Choosing 2% milk instead of whole milk reduces saturated fat from 5 g per cup to 3.5 g, a 30% reduction. In comparison, consuming 1% milk cuts the amount of saturated fat by 70% per cup.

Federal food and nutrition education programs promote adoption of the recommendations of the Dietary Guidelines for Americans [3]. Given the longstanding and consistent recommendation to consume 1% or nonfat milk, what reasons do consumers have for the slow adoption of low-fat milk? The major goal of this study is to report the spontaneous reactions of people who received one or more posts on their Facebook timeline during a five-week social media campaign on Facebook to promote consumption of 1% milk based on a content analysis of comments made by the public to the advertised posts. To our knowledge, this is the first study to report a content analysis of a social media nutrition education intervention. The reactions illustrate the competing perspectives toward the different types of milk and the utility of social media as a health promotion tool.

Intervention

The Oklahoma Nutrition Information and Education (ONIE) Project, a Supplemental Nutrition Assistance Program Education (SNAP-Ed) social marketing project, is funded by the United States Department of Agriculture through the Oklahoma Department of Human Services. As a SNAP-Ed grantee, the ONIE Project promotes good nutrition and physical activity, with a low-income priority audience (<130% federal poverty level), especially among those who are eligible for the Supplemental Nutrition Assistance Program (SNAP), formerly known as the Food Stamp Program.

SNAP is the largest food assistance program in the US. In 2014, Oklahoma had the 15th highest rate of SNAP participation in the US with a monthly average of 608,492 SNAP beneficiaries [10], or approximately one in six Oklahomans. SNAP recipients are free to choose what type of milk they will consume or to forego milk in favor of less healthy options such as sugar-sweetened beverages. Providing the SNAP population with nutrition information about healthy options consistent with the Dietary Guidelines for Americans [3] is the purpose of the SNAP-Ed program. The large number of SNAP recipients in Oklahoma gave the ONIE Project a sizable low-income audience for the social marketing intervention.

The purpose of this intervention, *Choose 1% Milk*, was to move higher-fat milk consumers to low-fat milk use. This multi-media social marketing intervention was the second of two interventions promoting low-fat milk use conducted by ONIE. The first, implemented in 2012, featured a local athletic celebrity and was implemented in the Oklahoma City media market. Using a quasi-experimental design, both pre- and post-intervention telephone surveys and milk sales data revealed a significant and positive change in 1% milk use in the intervention media market during the course of the intervention. [11,12]. The *Choose 1% Milk* intervention was implemented statewide over a five-week period. Social media, particularly Facebook, was a primary channel. Evaluation of milk sales data collected from 105 supermarkets located throughout the state of Oklahoma revealed that 1% milk sales increased significantly during the *Choose 1% Milk* intervention. The market share of 1% milk sold increased from 7.1% the week prior to the intervention to 10.1% immediately after it ended, a 42.9% relative increase. The overall quantity of milk sold did not significantly increase, suggesting that the increase in 1% milk sales was a result of high-fat milk users changing to 1% milk. In addition, nonfat milk sales did not significantly change, ruling out that nonfat milk users contributed to the increase in 1% milk sales.

2. Materials and Methods

2.1. Materials

The campaign logo was the image of an arm reaching for a gallon of 1% milk, with the word "Choose" prominently displayed. The tag line was "Choose 1% Milk", and the secondary tag line was "A Healthy Family Choice" (see Figure 2). The emphasis on choice came from studies of political communication by Luntz [13], and our own formative research that showed a positive reaction to messages conferring autonomy [14]. Similar to our previous intervention, the campaigns' key messages sought to dispel the most common myths held by SNAP recipients about 1% milk [14] including the differences in the fat content between 2% and 1% milk, that 1% milk has the same vitamins and minerals as whole milk, 1% milk is not watered-down, and 1% milk is recommended for children age two and older. All advertisements posted on social media relied heavily on infographic images. These images were simple but clever, and key messages communicated nutrition information with only a few words.

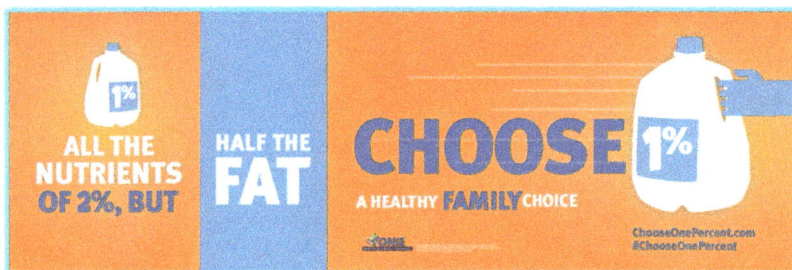

Figure 2. Choose 1% Milk Social Media Infographic.

2.2. Participants

ONIE collected the data from Facebook users who chose to respond to the health promotion messages about 1% milk during the *Choose 1% Milk* intervention by writing a comment about the post. There were no recruitment efforts or solicitations to obtain comments beyond the appearance of the paid advertisement (post) on each individual's timeline. In addition to residing in Oklahoma and adults age 18 and over, the characteristics of the individuals who would receive the post were specified when the advertisement was published on Facebook. Table 1 lists the priority audience profile for each post. Each post was targeted to the Facebook users who fit one or more of the characteristics specified. No personal information was gathered, so how many were SNAP participants is not known but there is no evidence to suggest that the SNAP population is underrepresented on Facebook.

Table 1. Facebook Advertisements and Social Media Participation (2014).

Post #	Date Posted	Advertisement	People Reached	Post-Clicks	Likes	Shares	Comments Analyzed	Cost Per Engagement (Like, Comment, Share & Post-Click)
	9/16	Campaign launched.						
1	9/16	Fact: 1% milk has all the nutrients of 2%, but with half the fat. Priority Audience: Motherhood, fatherhood, parenting, parent, pregnancy, family and relationships	110,592	1042	222	20	30	$0.44
2	9/18	Add a little sunshine to your diet with 1% low-fat milk. It has all the Vitamin D of 2% milk. Priority Audience: Fans and friends of fans	52,608	157	72	4	5	$1.24
3	9/19	Don't be a bonehead! Next time you buy milk, choose 1%! Priority Audience: Motherhood, fatherhood, parenting, pregnancy, family, and relationships	73,888	2103	1387	69	64	$0.16
4	9/21	Did you know that 1% milk has all the vitamins and minerals as 2% milk, but with half the fat? http://chooseonepercent.com Priority Audience: Motherhood, fatherhood, parenting, pregnancy, family, and relationships	99,424	745	655	29	38	$0.40
5	9/24	Build strong muscles and bones with 1% low-fat milk. It's the ultimate protein drink! http://chooseonepercent.com Priority Audience: Motherhood, fatherhood, parenting, pregnancy, family, and relationships	99,488	513	500	19	25	$0.70
6	9/26	We all know 1% milk has less fat, but does it have less vitamin D? No! Priority Audience: Motherhood, fatherhood, parenting, pregnancy, family, and relationships	110,208	974	1081	62	41	$0.34
7	9/26	Fact: 1% milk is not watered down! It has all the calcium of 2%, but less saturated fat. http://chooseonepercent.com Priority Audience: Fans and friends of fans	51,344	366	150	11	9	$1.15
8	9/30	Just the facts: 1% milk has all the protein, vitamin D and calcium of 2%, but with half the fat! Make the choice for a healthier you—and a healthier family. http://chooseonepercent.com Priority Audience: Low-income, parenting, family, and fitness	28,864	480	124	8	21	$0.82
9	10/7	FACT: 2% milk contains double the amount of fat found in 1%. Priority Audience: 18–65+	122,816	610	155	19	53	$0.67
10	10/7	Q: Which has more sugar, 1% milk or 2%? A: Trick question! They're the SAME. Priority Audience: Parenting	120,704	674	333	11	23	$0.95
11	10/8	Lots of people believe the fat in milk is healthy, but the majority of studies show that 1% is a better choice. Priority Audience: Fans and friends of fans	68,640	699	176	14	44	$0.99

Table 1. *Cont.*

Post #	Date Posted	Advertisement	People Reached	Post-Clicks	Likes	Shares	Comments Analyzed	Cost Per Engagement (Like, Comment, Share & Post-Click)
12	10/10	FACT: Once children reach 2 years of age, they don't need the fat in 2% and whole milk to grow. 1% is the healthy choice! Priority Audience: Mother, fatherhood, and parenting	147,904	1053	750	92	103	$0.55
13	10/13	On average, it takes 2 weeks to adjust to the taste of 1% milk and you still get the nine essential nutrients of milk! Priority Audience: Low-income	79,648	1077	220	23	51	$0.81
14	10/14	Get your fats straight! Priority Audience: 18–65+	161,792	829	312	12	47	$0.87
15	10/14	Kick the excessive sugars and additives in sports drinks. 1% milk is packed with vitamins and nutrients your body needs! Priority Audience: Parenting	102,112	402	491	25	18	$1.20
16	10/17	Children need milk. The American Academy of Pediatrics recommends 1% milk after the child's second birthday Priority Audience: Motherhood, fatherhood, parenting, pregnancy, family, and relationships	105,920	351	517	34	20	$0.57
	10/17	Campaign ended						
		Totals	1,535,952	12,075	7145	452	592	$0.74

2.3. Statistical Analysis

The themes of Facebook comments were identified qualitatively by four raters, each reading the comments independently. Each rater read the comments repeatedly to become familiar with them and to identify themes. The themes were not mutually exclusive, as raters may have assigned one comment to more than one theme.

In addition, the qualitative approach was supplemented by the quantitative method of cluster analysis. The need for cluster analysis comes from the fact that the social media material lacks the context of a group conversation, such as occurs with a focus group. Rather, social media content is briefer and less contextual. Cluster analysis has been in regular use in the social sciences since the 1960s and new applications continue to be reported [15–17]. We introduce cluster analysis as a quantitative method to classify the content of the themes identified by the raters for each comment. For this study, the analysis was a cluster of cases. Each case was a theme identified by an individual rater and groups of similar themes were identified by cluster analysis [18–20].

To provide converging evidence, two measures of similarity were used. The first measure of similarity was the Dice coefficient. It can range from 0 to 1, with a higher score meaning more similarity [21]. The second measure of similarity was Yule's Q, a correlation-like measure that ranges from −1 to +1, with a score of zero meaning no relationship [22]. A key property of Yule's Q is that when one rater's theme consists of a subset of comments from another rater's theme, then Yule's Q is equal to one, a relationship described as *complete* [23]. When Yule's Q is higher than the Dice coefficient, this suggests that one theme may be more general than another, so inspection of the items is warranted to explore this possibility.

For both the Dice coefficient and Yule's Q, themes were combined by average linkage between-groups. The final step in the analysis was again qualitative. The clusters of themes were examined qualitatively to verify similarities, and the raters reviewed the comments one last time to detect any themes that had not previously been identified.

3. Results

As seen in Table 1, during the *Choose 1% Milk* campaign, ONIE posted 16 different advertisements on Facebook. These advertisements were seen by 1.54 million people (reach). Reach is tracked by individual advertisment post not by the entire campaign, so this figure includes duplicated viewers who saw more than one of the posts. On average, each advertisement reached approximately 96,000 people (unique count). The total population in Oklahoma in 2014 was 3,878,051 people, meaning each advertisement was seen by 2.5% of the population [24]. A total of 8374 Facebook users engaged with the 16 posts by responding with a like, a share, or a comment. The 16 posts resulted in more than 12,000 clicks to read more about the topic, more than 7000 likes, and 452 shares. By definition, likes are positive endorsements of the message. A share indicates that the Facebook user shared the post on their timeline, which extends its reach to their family and friends. Another level of engagement is to write a comment, which may support or oppose the health message.

During the campaign, 454 Facebook users commented 592 times. Most commented only once (81.7%) or twice (12.8%), but 25 users (5.5%) commented three or more times. Of the 592 comments analyzed 103 (17.4%) were off topic, such as a thread in which users discuss diabetes with no reference to milk. Thus, there were 489 comments that related in some way to the health message about 1% milk.

The four raters individually read through these 489 comments and identified themes. The four raters identified 15, 18, 20, and 24 themes respectively for a total of 77 themes, some of which used very similar or the same words to describe the theme. These themes were subjected to cluster analysis so that themes sharing the same comments were grouped into similar clusters. Examination of the dendrogram suggested that a good solution was six broad thematic groups. A total of 19 more specific sub-themes emerged from within the broad thematic groups as summarized below in Table 2. In the final qualitative step, raters reviewed the comments and themes, and no additional themes were identified. Discussion of the themes is ordered by the number of comments in each group, presuming that frequency is indicative of more common or more intensely held beliefs.

Table 2. Themes and Sub-themes of the Content Analysis.

Thematic Group	Sub-Theme *Comment [Facebook Post # in Table 1]*
	Fat and Sugar ($n = 37$) *What they take out of whole milk they replace with sugar/sweetener. [#5]* *The lower the fat content, the higher the sugar content. [#1]* *And full of sugar! [#7]* *Flavored sugar filled milk contributes way more to childhood obesity. [#12]* *This is so wrong. Stop selling garbage information to young parents and killing our babies with sugar...[#12]* *Sugar water [#3]*
1. Sugar, Fat, and Nutrients ($n = 118$)	Nutrients ($n = 35$) *I could be wrong, but aren't all nutrients super-heated out then added back in? Hydrogenated [sic] means super-heated. [#13]* *1% has more vitamins because it has more milk. [#9]* *With that logic lots of things are "packed" with nutrients, but the amounts are very little. [#13]* *One percent milk has no vitamins to make bones grow strong and help the teeth also.[#12]*
	Healthy Fat ($n = 31$) *Actually low fat has been shown to be ineffective at weight loss. [#7]* *Fat is required to digest nutrients found in milk. [#14]* *Did you know that the fat in whole milk is good for you? [#4]* *Yup fat and cholesterol are good for you. [#1]* *Wow another lie this milk . . . has no vitamins at all bc it's in the fat content...take a nutrition class. [#2]*
	Fat is Good for Children ($n = 16$) *There are fats in whole milk that help children's brain development. [#14]* *For adults it's better, but for children it's not. [#6]* *Children's brains NEED good fats to develop. [#12]*

Table 2. *Cont.*

Thematic Group	Sub-Theme *Comment [Facebook Post # in Table 1]*
2. Defiant (*n* = 110)	Rejection (*n* = 75) *BS, 1% milk is worthless.* [#4] *No one really knows wtf is good or bad for you now-a-days.* [#3] *Don't care. Not drinking 1%.* [#10]
	Sarcasm (*n* = 30) *Well duh!* [#9] *How much time and money were wasted to come to this incredible conclusion?* [#9]
	Tyranny (*n* = 8) *How about we let the parents do the parenting......1% is yuk!* [#12] *Government at work in our homes now!!!!!* [#4] *Stupid government telling people how to eat and everyone's falling for it. SMH.* [#12]
3. Watery Milk (*n* = 98)	Sensory Experience (*n* = 77) *Might as well drink water. No taste.* [#6] *It's blue tinged.* [#12] *When I drink milk, I want to taste milk, not water.* [#9] *My words exactly whole milk...1% is blue water!!!* [#3]
	Watered Down (*n* = 31) *We come from a dairy farm and we don't want our milk watered down.* [#3] *100% milk builds strong bones. Not watered down fake 1%!* [#3] *It is mostly water.* [#4] *It's just water with a splash of milk in it duh* [#9]
	Cost (*n* = 8) *They make more money off 1% milk! (Less milk and more water!)* [#3] *It's about money.* [#10] *You are paying for mostly water.* [#13] *Read the labels folks. They must own stock in a water company* [#16]
4. Personal Preference (*n* = 93)	Likes High-Fat Milk (*n* = 50) *Whole milk all the way.* [#3] *Whole milk is my favorite.* [#5] *2% is my fav.* [#6]
	Likes Low-Fat Milk (*n* = 30) *I love skim milk!* [#1] *1% is our choice.* [#5] *We have been drinking 1% for years, and love it.* [#9]
	Raw Milk (*n* = 23) *Straight from the cow or nothing.* [#6] *Not just whole milk but whole RAW milk is the healthiest of all.* [#4] *I want it fresh from the cow.* [#8]
5. Evidence and Logic (*n* = 79)	Scientific Evidence (*n* = 38) *False advertising. There is absolutely no scientific evidence that people who drink cow's milk have stronger bones.* [#3] *Studies have come out to show that's not true. Whole milk is best.* [#12] *My son's Dr says that 2% is best.* [#12] *I challenge the person who came up with this crazy idea to prove it.* [#12]
	Anecdotal Evidence (*n* = 32) *Whole milk is the best. Pioneers didn't have 1% and they survived just fine.* [#3] *I find it funny that our grandparents and great grandparents drank whole milk if it wasn't straight from a cow and lived to be a ripe old age.* [#3] *I've been drinking whole milk my entire life and have never been over weight, even after two kids.* [#13]
	Exercise (*n* = 14) *Walk a little more and enjoy the 2 percent.* [#9] *Fat free or low fat isn't the answer. Portion control and exercise.* [#14] *Wow the Time we talk about this you could have walked outside and said high to the neighbors, and burnt more fat than a glass of milk would provide.* [#12]
6. Pure and Natural (*n* = 64)	Nature (*n* = 33) *There's nothing wrong with the milk God made.* [#12] *The vitamins and minerals are added to 1% they aren't natural.* [#9] *Milk is from nature. Store milk is from chemistry.* [#6]
	Milk Alternatives (*n* = 21) *I will just drink my almond milk!* [#14] *Cow's milk is nasty. I much prefer plant milks.* [#3] *Use unsweetened soy milk and let the calves drink the milk they should be getting.* [#13]
	Contamination (*n* = 19) *May have Vitamin D but it also has growth hormones that are not good for us!!* [#2] *Milk has pus from cows, hormones and antibiotics.* [#13] *Investigate the chemicals used to remove the fat from milk!!* [#6]

Note: The *n* for each major thematic group is the number of non-duplicate comments in the group, which may be less than the sum of the *n* comments for the sub-themes in that group.

3.1. Thematic Group 1: Sugar, Fat, and Nutrients

Sugar, fat, and nutrients had the most comments ($n = 118$) with four sub-themes. The comments in this thematic group, whether positive or negative, refer to beliefs about the nutrients in different types of milk. Comments in the sub-theme of *Fat and Sugar* ($n = 37$) refer to sugar or fat or both. Some comments were neutral, but many contained arguments against drinking 1% milk. Some Facebook users mistakenly believed that sugar was added to low-fat milk. Others considered 1% milk a trade-off, such that less fat means more sugar.

Comments in the sub-theme of *Nutrients* ($n = 35$) refer to the general topic of nutrition, such as the mention of calcium or protein, or of individual differences in nutritional needs. Generally, these comments claimed that the vitamins and minerals in low-fat milk are lower in amount or quality. A second erroneous assumption was that all the vitamins in milk are fat soluble, so drinkers of low-fat milk do not absorb the vitamins in milk.

In the sub-theme of *We Need the Fat*, there are 31 comments. These comments argue that whole milk is better, because dietary fat is healthy or at least not harmful to health. Also, like the comments in the *Nutrients* sub-theme, there is the belief that low-fat milk is less nutritious. However, the comments in this sub-theme imply that the nutrients in milk are in the fat, and that removing the fat also removes the nutrients.

Finally, in the sub-theme of *Fat is Good for Children*, there are 16 comments. These comments refer specifically to the nutritional needs of children. These comments support giving whole milk to children, and many express the belief that fat is important for children age two and older to develop; in fact, about one out of three of these comments specifically referred to brain development.

3.2. Thematic Group 2: Defiant

Harsh or negative comments rejecting the health promotion message cluster into the second largest thematic group ($n = 110$) with three sub-themes. The 75 comments in the sub-theme *Rejection* disputed the message. Examples were "why bother", "no way", "bull" and "yuck." Some comments were hostile or contained profanity, and others questioned the validity of the key messages.

The sub-theme labeled *Sarcasm* consisted of 30 comments ridiculing the post. Notably, half of these comments emerged in response to one advertisement (Table 1, #9) that illustrated the difference in the fat content between 2% and 1% milk with the key message, "FACT: 2% milk contains double the amount of fat found in 1%". Responses included "Well duh!" and "How much time and money were wasted to come to this incredible conclusion?" This particular ad received few likes, as well.

The third sub-theme has only eight comments, and it is labeled *Tyranny*. These comments often explicitly say that the health message was wrong. For example, "This is all hog wash! Give those kids REAL milk and cut the fat somewhere else." They also refer to the message as an example of tyrannical political action and the government telling people what to do.

3.3. Thematic Group 3: Watery Milk

The formative research had revealed a pervasive belief that 1% milk was watered-down whole milk, and comments in this theme affirm that this is an important belief influencing the type of milk consumed. Each subtheme captures a unique aspect of this belief.

The *Sensory Experience* sub-theme made invidious comparisons of 1% to higher-fat milk, stating that whole milk is an appealing white with a pleasing thick texture and rich taste, but that 1% milk has a bluish tint with a watery texture and no flavor ($n = 31$). In contrast, comments in the sub-theme named *Watered Down* ($n = 33$) explicitly spoke of added water or called 1% milk white water or colored water. *Cost*, the last sub-theme in this group, contained eight comments. These comments express the view that milk producers and marketers added water to milk to make more money.

3.4. Thematic Group 4: Personal Preference

The comments clustered together in this theme focused on what type of milk the commenter preferred and not the attributes of the type of milk (n = 93). Slightly more than half of these comments expressed a personal liking for 2% or whole milk through attitude or behavior. The 30 comments in the sub-theme *Likes Low-Fat Milk* are similar, but the personal preference is for low-fat milk. Finally, in a third sub-theme named *Raw Milk*, the Facebook user stated the desire to have raw milk, or to have milk straight from the cow (n = 33).

3.5. Thematic Group 5: Evidence

Comments of the Facebook users in this theme (n = 79) offered scientific, antidotal, or well-established facts to justify their beliefs. Those in the sub-theme we labeled *Scientific Evidence* (n = 38) justified their position by referencing scientific evidence. The source may have been rather vague, such as a reference to research studies in general, or the source may have been fairly specific, as in a well-known television personality who offers expert advice, or their family physician. Some Facebook users cited a media outlet as the source of evidence or provided links to research findings from the media outlet itself.

In the sub-theme named *Anecdotal Evidence* (n = 32) comments cited a personal experience or anecdotal evidence, not scientific research. These Facebook users usually referred to themselves or family members who have used whole milk for years and are healthy. Some people referenced ancestors' consumption of milk before the west was settled, presuming these pioneers had been healthy.

The sub-theme named *Exercise* has 14 comments. The nature of the evidence in these comments is the fact that weight can be lost by increasing physical activity. These comments rejected the message of the health benefits of low-fat milk by asserting that the problem can be resolved with exercise.

3.6. Thematic Group 6: Pure and Natural

The last thematic group contained 64 comments, which was the smallest number of comments (13.1% of the total). These comments rejected the use of 1% milk on the grounds that it is not pure or natural.

Comments grouped in the sub-theme *Nature* (n = 33) reflected a belief that adult milk consumption was unnatural, noting that milk is for babies and not for adults. Others rejected milk sold in stores on the grounds it was an unnatural processed food item, preferring raw milk. Similarly, but slightly different, others rejected low-fat milk claiming that it is processed but whole milk is natural.

There were 21 comments in the sub-theme *Milk Alternatives*. These statements rejected the use of dairy milk, sometimes on the grounds that drinking the milk of another species is unnatural, and recommend the use of milk from plants sources, such as almond or soymilk.

The third sub-theme in this group is labeled *Contamination*. The 19 comments in this sub-theme refer to the belief that milk products are contaminated. In some comments, the nature of the contaminant vaguely referenced chemicals or additives. Others were more specific, as in a reference to hormones, antibiotics, pus, or pesticides. A number of Facebook users linked their comments to blogs that assert that dairy contains a host of contaminates and is not safe for human consumption as evidence supporting their comments. These commenters were strongly opposed to the use of dairy milk.

4. Discussion

Facebook is a nascent community of sorts and some users want to actively participate or interact with others. A well-executed social media campaign captures the attention of the audience, who then engage with both negative and positive comments. Yet, public health campaigns have been slow to make use of the interactive nature of social media [25]. This study examined the content of spontaneous

Facebook comments made in response to a successful social marketing intervention with statewide exposure on social media, the *Choose 1% Milk* campaign.

Six broad themes illustrate the types of comments that emerged. These included: (1) *Sugar, Fat, and Nutrients,* the interest in the nutritional content of milk; (2) *Defiant,* the expression of defiant and negative attitudes toward the health message; (3) *Watery Milk,* the concern that low-fat milk has added water or has the sensory qualities of water; (4) *Personal Preference,* a statement endorsing what is familiar and liked in a milk product; (5) *Evidence and Logic,* the dismissal of the health benefits of 1% milk by appeal to contrary anecdotes or to the lack of research support for the touted benefits; and (6) *Pure and Natural,* the idea that whole milk is pure and natural, but low-fat milk is impure, adulterated, or unnatural.

The thematic group *Watery Milk* contained the most common sub-theme *Sensory Experience,* with 77 comments about flavor and texture. This theme was expected, because concerns about the sensory difference between whole milk and low-fat milk appeared in our formative research, and because sensory preferences are a recognized motive for food choice [26]. In fact, taste preference is a leading reason to choose food items [26,27].

There was also the theme *Watered Down,* whose comments expressed the belief that low-fat milk has added water, implying inferior nutrition. Again, we expected to encounter this belief, as it appeared in the formative research [12,14], and it has appeared in previous research on low-fat milk [28]. These comments affirm the relevance of key messages and the need to repeat messages challenging strongly held beliefs to create a shift in attitudes [29]. Moreover, the interactive nature of social media is a desirable feature that allowed the campaign team to reply with comments to correct this mistaken idea of added water. A possible strategy for future campaigns is to monitor the frequency of such comments over time to see if they diminish.

The benefit of using the interactive nature of social media is illustrated with the thematic group *Sugar, Fat, and Nutrients.* One sub-theme in this group was *Fat and Sugar,* the belief that lower fat in milk means more sugar. This belief did not appear in formative research but emerged as a barrier to the adoption of low-fat milk for this audience segment. Because Facebook is not static, new advertising content can be developed quickly and inexpensively. ONIE took advantage of this feature and posted: "Q: Which has more sugar, 1% milk or 2%? A: Trick question! They're the SAME." This advertisement reached 120,704 Facebook users and received a high level of engagement.

Effective use of social media's ability to react quickly and interact directly with the audience is also illustrated with the thematic group *Pure and Natural.* Some Facebook users argued that milk from soy or almonds was more natural, using the *Choose 1% Milk* campaign to promote their personal philosophical nutrition beliefs. These non-dairy milk users were not the priority audience for *Choose 1% Milk* campaign. Yet, comments devaluing dairy milk had the potential to adversely affect milk consumption, an unintended and undesirable outcome. The ONIE team politely responded to some of these comments to reassure milk users that dairy milk is safe. The lesson for public health practitioners is that in an interactive campaign on any topic, one may be called on to address related competing behaviors and other potential unintended consequences.

Comments in the thematic group *Personal Preference* referred to milk that was familiar and typically consumed. This theme is similar to the familiarity motive from the Food Choice Questionnaire [26], in which people said they chose a food because it "is what I usually eat" or "is like the food I ate when I was a child." ONIE formative research also identified this theme [12,14]. As its strategy to reduce familiarity as a barrier to low-fat milk use, when online comments promoted whole or 2% milk, the ONIE team offered no response; but when comments supported low-fat milk, ONIE "liked" the comments and crafted responses affirming the behavior. In this way, the interactive nature of social media was used to further promote a positive health message.

Another thematic group in the Facebook comments was called *Evidence and Logic.* This audience segment claims to select food to promote health. Regardless of the type of milk consumed, milk users often identified health benefits as a motivating factor for the type of milk used [12,14], a factor identified as influencing food choices generally as well [26]. This audience segment wants current scientific evidence of the health benefit and resolution of conflicting nutritional advice. Previous research has found a desire by mothers in a nutrition program to be provided with information based on trustworthy research, including external links to more information [30], and the Food Hero program recommends providing research-based information [31]. To support these Facebook users need for more information, ONIE project staff responded with research-based information and links to more information. A strategy for future campaigns would be to compile a summary of credible research findings and share the summary in replies to those who seem to want evidence to support particular claims. Also, comments directed to ONIE asked who we were and what our motives were, so we were transparent about our funding source and mission.

At the same time, not all of the comments indicated openness to scientific evidence. The thematic group *Evidence and Logic* included the theme of *Anecdotal Evidence*. These comments put little credence in scientific research evidence, and instead appealed to anecdotes. The study of the spread of misinformation has been described as network sociology, and "it may be as influential as the information content and scientific validity of a particular health topic discussed using social media" ([32], p. 517). It may be useful for future research to examine how an appeal to anecdotes affects the spread of public health messages.

The last thematic group was named *Defiant*. These comments expressed negative attitudes using sarcasm, hostile remarks, profanity, and cynical attitudes toward government, many of which were not appropriate for a family-oriented Facebook page. Comments like these did not arise during the formative research from focus groups or from a telephone survey, so the negative comments appear to arise from the nature of social media. One challenge for social media campaigns is how to address such negative comments. One study of Twitter examined tweets on the public health topic of HPV vaccines, with the results indicating that users exposed to negative tweets were more likely to make a negative tweet [33]. To manage the social media, ONIE staff monitored comments throughout the day and evening. Profanity was deleted, and gratuitous insults were hidden from view of anyone except ONIE staff and the Facebook user making the comment. Comments that were merely negative about the health message, with no profanity or degrading language, were ignored. It can be noted that many of these comments were directed at one post (Table 1, #9). Perhaps this message was interpreted as condescending, obvious by definition, or too simple to be interesting or informative. So there may be a lesson to be learned for social media campaigns on any health topic: messages must be simple and direct but without appearing condescending.

We found the cluster analytic method introduced here to be a useful tool to study social media content. Unlike material from a focus group in which statements are part of a dialogue and comments are related to what comes before as well as after the comment (i.e., have a meaning-laden context), the Facebook comments were brief statements, usually one sentence, and sometimes merely a phrase. Material from other social media platforms is similarly stand-alone content, such as tweets, or pictures from Instagram. Given the lack of context and narrative structure to the material and the depersonalized context, cluster analysis could be helpful in identifying similar themes. The use of and consensus among four raters also added evidence for the reliability of the results.

To maximize the use of the interactive nature of social media, we have several suggestions: be transparent about funding sources and the purpose of the campaign; repeat key messages in advertising posts countering or challenging strongly held beliefs and monitor comments to identify whether the frequency of these type of comments diminish; keep comments brief and simple without appearing condescending; develop new posts to respond quickly to themes that emerge from comments; dedicate staff to monitor comments, delete or hide those with profanity or personal attacks, engage with Facebook users who comment, and assess the influence of comments on the promoted behavior. Another lesson learned is to promote engagement by directly asking the audience to participate such as by asking a question, which allows viewers to participate immediately and quickly.

Beverages **2017**, 3, 47

Social media is quickly becoming a dominant source of all communication including health communication, transforming with whom and how people interact, obtain information and voice their opinions. It is more engaging than other media channels for health promotion interventions because it supports dialogue, a key element in developing a critical perspective on a topic [34]. In addition, it is a sure pathway into the social networks of a person's friends and family. Given the rapid extension of social media, public health needs to embrace this medium and do everything to add to our understanding of how to make it work to improve population health.

The adoption of 1% milk has been slow in the US. Americans need a compelling reason to adopt 1% low-fat milk. If adopting 1% low-fat milk provides health benefits sufficient to warrant repeated recommendation in the last six Dietary Guidelines for Americans, what can be done to improve adoption by the public? The comments received to the 16 posts, all of which were consistent with the recommendations made in the Dietary Guidelines for Americans since 1990, provide additional insight on the question. The subject of milk is a contested terrain in the mind of the consumer with a variety of competing perspectives grounded in accurate or mistaken nutrition knowledge about the nutritional content of the different types of milk, rational and non-rational thought (scientific evidence vs. personal preference, feelings about family or personal experience), competing scientific findings (good fat, bad fat), and political philosophy and beliefs about nature, government tyranny and corporate greed. Although some of the reasons to reject any change in milk consumption could be anticipated (such as lack of or mistaken nutrition knowledge or preference for a non-animal diet), milk consumption is a controversial issue and it is easy for people to find an opinion about milk that makes sense to them. If adopting 1% milk has all of the alleged or anticipated benefits, making a more explicit recommendation every five years will not change behavior very quickly as evidence from the last 27 years reveals. Instead, a concerted and consistent effort to acknowledge all of these competing claims and address them to the extent possible is the only apparent way to achieve the recommendation to make consumption of 1% or nonfat milk the social norm.

Acknowledgments: Partial support for this study came from the US Department of Agriculture through the Oklahoma Department of Human Services through a Supplemental Nutrition Assistance Program, Nutrition Education Grant (Robert John, PI). We would like to Meredith S. Scott, M.S., Project Coordinator for ONIE, for her contributions to the design and execution of the intervention. This research was a component of a 1% low-fat milk intervention that received the National Social Marketing Centre (Great Britain) Award for Excellence in Social Marketing announced at the University of South Florida's 24th Social Marketing Conference, 2016. The funding agencies had no role in the design, analysis, or writing of this article. The interpretations expressed here are the authors and do not reflect the views of the funding agencies.

Author Contributions: All authors listed contributed substantially to this project and manuscript. R.J., K.J.F., and D.S.K. conceived and designed the study; K.J.F. assembled the data; D.S.K. performed the quantitative data analysis; R.J., D.S.K., K.J.F., J.O. and K.H. analyzed the qualitative data and contributed to the interpretation; R.J., K.J.F., and D.S.K. made substantial contributions to writing the manuscript; and all named authors have reviewed the manuscript.

Conflicts of Interest: The authors declare no conflict of interest. The funding agencies had no role in the design of the study, in the collection, analysis, or interpretation of data, or the writing of the manuscript.

References

1. US Department of Agriculture and the US Department of Health and Human Services. *Scientific Report of the 2015 Dietary Guidelines Advisory Committee*; US Human Nutrition Information Service: Washington, DC, USA, 2015.

2. US Department of Agriculture and the US Department of Health and Human Services. *Nutrition and Your Health: Dietary Guidelines for Americans*, 3rd ed.; US Human Nutrition Information Service: Washington, DC, USA, 1990. Available online: https://health.gov/dietaryguidelines/1990.asp (accessed on 12 April 2017).

3. US Department of Health and Human Services and US Department of Agriculture. *2015–2020 Dietary Guidelines for Americans*, 8th ed.; Department of Health and Human Services: Washington, DC, USA, 2015. Available online: http://health.gov/dietaryguidelines/2015/guidelines/ (accessed on 1 February 2016).

4. US Department of Agriculture Services and US Department of Agriculture. *Dietary Guidelines for Americans*, 7th ed.; Government Printing Office: Washington, DC, USA, 2010. Available online: https://health.gov/dietaryguidelines/dga2010/dietaryguidelines2010.pdf (accessed on 20 November 2011).
5. US Department of Agriculture Services and US Department of Agriculture. *Dietary Guidelines for Americans*, 6th ed.; Government Printing Office: Washington, DC, USA, 2005. Available online: https://health.gov/dietaryguidelines/dga2005/document/ (accessed on 12 April 2017).
6. US Department of Agriculture and the US Department of Health and Human Services. *Nutrition and Your Health: Dietary Guidelines for Americans*, 5th ed.; Government Printing Office: Washington, DC, USA, 2000. Available online: https://health.gov/dietaryguidelines/2000.asp (accessed on 12 April 2017).
7. US Department of Agriculture and the US Department of Health and Human Services. *Nutrition and Your Health: Dietary Guidelines for Americans*, 4th ed.; Government Printing Office: Washington, DC, USA, 1995. Available online: https://health.gov/dietaryguidelines/1995.asp (accessed on 12 April 2017).
8. US Department of Agriculture. In Fluid milk sales by product (Annual) [Data file]; 2012. Available online: http://www.ers.usda.gov/data-products/dairy-data.aspx (accessed on 5 May 2017).
9. Rehm, C.D.; Drewnowski, A.; Monsivais, P. Potential Population-level Nutritional Impact of Replacing Whole and Reduced-fat Milk with Low-Fat and Skim Milk among US Children Aged 2–19 years. *J. Nutr. Educ. Behav.* **2015**, *47*, 61–68. [CrossRef] [PubMed]
10. US Department of Agriculture Food and Nutrition Service. (2015) Supplemental Nutrition Assistance Program State Activity Report: Fiscal Year; 2014. Available online: www.fns.usda.gov/sites/default/files/FY14%20State%20Activity%20Report.pdf (accessed on 4 April 2017).
11. Finnell, K.J.; John, R.; Thompson, D.M. 1% Low-Fat Milk has Perks!: An Evaluation of a Social Marketing Intervention. *Prev. Med. Rep.* **2017**, *5*, 144–149. [CrossRef] [PubMed]
12. Finnell, K.J.; John, R. A Social Marketing Approach to 1% Milk Use: Resonance is the Key. *Health Promot. Pract.* **2017**. [CrossRef] [PubMed]
13. Luntz, F. *Words That Work: It is Not What You Say, It is What People Hear*; Hyperion: New York, NY, USA, 2007; ISBN-13: 978-1401302597.
14. Finnell, K.J.; John, R. Formative Research to Understand the Psychographics of 1% Milk Consumption in a Low-Income Audience. *Soc. Mark. Q.* **2017**, *23*, 169–184. [CrossRef]
15. Blashfield, R.K. Mixture Model Tests of Cluster Analysis: Accuracy of 4 Agglomerative Hierarchical Methods. *Psychol. Bull.* **1976**, *83*, 377–388. [CrossRef]
16. Johnson, S.C. Hierarchical Clustering Schemes. *Psychometrika* **1967**, *32*, 241–254. [CrossRef] [PubMed]
17. Kettenring, J.R. The Practice of Cluster Analysis. *J. Classif.* **2006**, *23*, 3–30. [CrossRef]
18. Borgen, F.H.; Barnett, D.C. Applying Cluster Analysis in Counseling Psychology Research. *J. Couns. Psychol.* **1987**, *34*, 456–468. [CrossRef]
19. Clatworthy, J.; Buick, D.; Hankins, M.; Weinman, J.; Home, R. The Use and Reporting of Cluster Analysis in Health Psychology: A Review. *Br. J. Health Psychol.* **2005**, *10*, 329–358. [CrossRef] [PubMed]
20. Punj, G.; Stewart, D.W. Cluster Analysis in Marketing Research: Review and Suggestions for Application. *J. Mark. Res.* **1983**, *20*, 134–148. [CrossRef]
21. Dice, L.R. Measures of the Amount of Ecologic Association between Species. *Ecology* **1945**, *26*, 297–302. [CrossRef]
22. Yule, G.U. On the Methods of Measuring Association between Two Attributes. *J. R. Stat. Soc.* **1912**, *75*, 79–652. [CrossRef]
23. Kendall, M.G.; Stuart, A. Inference and Relationship. In *The Advanced Theory of Statistics*; Griffin: London, UK, 1961; Volume 2.
24. U.S. Census Bureau. American Community Survey-1 Year Estimate: Summary File B01003: Total Population; 2014. Available online: https://factfinder.census.gov/ (accessed on 30 June 2017).
25. Thackeray, R.; Neiger, B.L.; Smith, A.K.; Van Wagenen, S.B. Adoption and Use of Social Media among Public Health Departments. *BMC Public Health* **2012**, *12*, 242. [CrossRef] [PubMed]
26. Steptoe, A.; Pollard, T.M.; Wardle, J. Development of a Measure of the Motives Underlying the Selection of Food—The Food Choice Questionnaire. *Appetite* **1995**, *25*, 267–284. [CrossRef] [PubMed]
27. John, R.; Kerby, D.S.; Landers, P.S. A Market Segmentation Approach to Nutrition Education among Low-Income Individuals. *Soc. Mark. Q.* **2004**, *10*, 24–38. [CrossRef]

28. Wechsler, H.; Wernick, S.M. A Social Marketing Campaign to Promote Low-Fat Milk Consumption in an Inner-City Latino Community. *Public Health Rep.* **1992**, *107*, 202–207. [PubMed]

29. Lee, N.R.; Kotler, P.A. *Social Marketing: Changing Behaviors for Good*, 5th ed.; SAGE Publications: Thousand Oaks, CA, USA, 2015; ISBN-13: 978–1452292144.

30. Leak, T.M.; Benavente, L.; Goodel, L.S.; Lassiter, A.; Jones, L.; Bowen, S. EFNEP Graduates' Perspectives on Social Media to Supplement Nutrition Education: Focus Group Findings from Active Users. *J. Nutr. Educ. Behav.* **2014**, *46*, 203–208. [CrossRef] [PubMed]

31. Tobey, L.N.; Manore, M.M. Social Media and Nutrition Education: The Food Hero Experience. *J. Nutr. Educ. Behav.* **2014**, *46*, 128–133. [CrossRef] [PubMed]

32. Seymour, B.; Getman, R.; Saraf, A.; Zhang, L.H.; Kalenderian, E. When Advocacy Obscures Accuracy Online: Digital Pandemics of Public Health Misinformation through an Antifluoride Case Study. *Am. J. Public Health* **2015**, *105*, 517–523. [CrossRef] [PubMed]

33. Dunn, A.G.; Leask, J.; Zhou, X.J.; Mandl, K.D.; Coiera, E. Associations between Exposure to and Expression of Negative Opinions about Human Papillomavirus Vaccines on Social Media: An Observational Study. *J. Med. Internet Res.* **2015**, *17*, 10. [CrossRef] [PubMed]

34. Freire, P. *Pedagogy of the Oppressed*, 1st ed.; The Seabury Press: New York, NY, USA, 1970; ISBN-13: 978-0826412768.

beverages

MDPI

Article

Research to Understand Milk Consumption Behaviors in a Food-Insecure Low-Income SNAP Population in the US

Karla Jaye Finnell and Robert John *

Health Sciences Center, University of Oklahoma, Oklahoma City, OK 73126-0901, USA; karla-finnell@ouhsc.edu
* Correspondence: robert-john@ouhsc.edu; Tel.: +1-405-271-2017 (ext. 46755)

Academic Editor: Alessandra Durazzo
Received: 13 July 2017; Accepted: 25 August 2017; Published: 18 September 2017

Abstract: Milk, due to its affordability and nutritional value, can fortify the diets of families that experience food insecurity or find a high-quality diet cost-prohibitive. However, it can also be a leading source of excess calories and saturated fat. Yet, little is known about what influences consumer behavior of Supplemental Nutrition Assistance Program (SNAP) recipients toward the type of milk used or the prevalence of low-fat milk use among this population. This cross-sectional telephone survey of SNAP recipients (n = 520) documented that 7.5% of this population usually consumes low-fat milk, a prevalence that lags behind national figures (34.4%) for the same time-period. There was a weak association between sociodemographic characteristics of SNAP recipients and low-fat milk use. Instead, less low-fat milk consumption was associated with a knowledge gap and misperceptions of the nutritional properties of the different types of milk. Promoting low-fat milk use by correcting these misperceptions can improve the diet of America's low-income population and reduce food insecurity by maximizing the nutritional value of the foods consumed.

Keywords: milk; nutrition; food security; psychographics; knowledge; attitudes and beliefs

1. Introduction

Although the prevalence of food insecurity has declined as the economy has recovered from the Great Recession that began in December 2007, nearly 16 million US households experienced food insecurity at some point in 2015 [1]. This means that 12.7% of US households do not have the resources to ensure consistent and dependable access to a healthy diet. Most of these food insecure households (59%) participate in one or more of the federal nutrition assistance programs including the Supplemental Nutrition Assistance Program (SNAP), the National School Lunch Program, and the Special Supplemental Nutrition Program for Women, Infants, and Children (WIC). Of these, SNAP provides most of the nutrition support and SNAP-Ed agencies, an ancillary program, deliver free nutrition education for low-income households based on the Dietary Guidelines for Americans [2].

Milk, because it is affordable, can fortify the diets of families that experience food insecurity or find a high-quality diet cost-prohibitive [3–9]. It is rich in protein and Vitamin B12, as well as under-consumed nutrients such as Vitamin A, Vitamin D, potassium, calcium, and magnesium [3,4,6–8,10]. Research has shown that, on average, milk contributes 49.5% of Vitamin D, 25.3% of calcium, 17.1% of Vitamin B12, and 11.6% of potassium to the American diet [6,7]. On the other hand, milk is among the top ten sources of saturated fat and calories [6]. More specifically, based on data from the 2003–2006 National Health & Nutrition Examination Survey, milk was the third leading source of saturated fat and the seventh leading source of calories in the American diet [6].

Both saturated fat and excess calorie intake are linked to obesity, heart disease, and type 2 diabetes—increasing the burden of chronic disease [11]. Not surprisingly, these diseases occur more often among populations with limited financial resources. Research shows that adults with income below 130% of the federal poverty level that participate in SNAP are more likely to be overweight or obese and experience more heart disease, stroke, or diabetes than non-participants with higher incomes [11].

There are small, dietary changes that can significantly reduce the number of calories and the amount of saturated fat consumed without compromising nutritional intake. Nonfat and 1% have the nutritional benefits of high-fat milk but with less saturated fat and fewer calories [2]. In a study of children (age 2 to 19), researchers concluded that replacing whole and 2% milk with nonfat or 1% milk would significantly reduce the amount of saturated fat and calories consumed at a population level without compromising the nutrient intake of potassium or calcium [10].

Although not all researchers agree that consumption of saturated fat from dairy increases health risks [12–17], the Scientific Advisory Committee for the Dietary Guidelines continues to recommend that low- and nonfat dairy offers key nutrients, but with less saturated fat (a nutrient overconsumed by Americans) [18]. Subsequently and consistent with its past recommendations [19,20], the 2015–2020 Dietary Guidelines [2] reaffirmed the use of 1% or nonfat milk. These guidelines serve as the basis for federal food and nutrition education programs, such as SNAP-Ed.

Even though low-fat milk is recommended as part of a healthy diet, most Americans consume high-fat milk. The 2003–2004 National Health and Nutrition Examination Survey documented that 74.0% of all milk consumed was whole (32.3%) or 2% milk (41.7%); 1% and nonfat milk together represented 26.0% of all milk consumed (10.4% and 15.6%, respectively) by the general population [21]. Similarly, low-fat milk sales represented 28.6% of all milk sales for this same time-period, and by 2012, low-fat milk represented barely more than a third of all milk sales (34.4%) [22].

Not only has low-fat milk use only marginally improved over the past decade, but those with lower household income and less educational attainment are also less likely to use low-fat milk [23–25]. In a study examining data from the 1996 National Food Stamp Program Survey, 15.6% of SNAP recipients reported using low-fat milk [26], a prevalence that is barely half that of the general population [22]. Low-fat milk use ranges from 38.1% among children and adolescents living in households with income greater than or equal to 350% of the federal poverty level (FPL) to a low of 9.4% among those living in households with income below 130% of the FPL (i.e., eligible to participate in SNAP) [24].

Americans age two and older consumed about three-fourths of an 8-oz cup of milk each day [8], which is substantially below the three 8-oz cups of low-fat milk that is recommended for daily consumption by the US Dietary Guidelines [2]. Moreover, the mean intake of milk gradually declines with age. That is, children consume the most milk, followed by adolescents, and then adults. Yet, unlike the type of milk, income has not been found to be strongly associated with the quantity of milk consumed except among adolescents. Adolescents living in households with an income greater than 350% of FPL consumed significantly more milk than those living in a household with income between 101% and 185% of the FPL [8].

Notwithstanding the documented disparity in low-fat milk use by socioeconomic status, no peer-reviewed studies have explored the knowledge, attitudes and beliefs of SNAP recipients toward low-fat milk. In one study, women with income higher than the median reported foregoing their taste preference for higher-fat milk to consume low-fat milk, which they perceived as healthier [27]. These findings were the basis of a series of *1% or Less* interventions conducted in the 1990s' in West Virginia [28–32], which were later replicated in California [33] and Hawaii [34]. Relying on data collected before and after one of the *1% or Less* campaigns [29], Butterfield and Reger concluded that, in part, positive changes in beliefs about the healthiness (less saturated-fat), taste, and cost of low-fat milk, led to a significant increase in use of low-fat milk [28].

Maglione and colleagues, building on the findings of Booth-Butterfield and Reger [28], explored attitudes associated with low-fat milk in Hawaii [25]. Using a 4-item scale to measure attitudes (taste and health benefits) and a 3-item scale to measure normative beliefs (community, doctor, and friends and family), low-fat milk users had more positive attitudes toward low-fat milk than high-fat milk users, as well as more positive normative beliefs. Applying the transtheoretical model of change to measure intent to use low-fat milk, these researchers posited an intervention promoting low-fat milk could have a population-level impact although intention to change to low-fat milk was modest. Other studies have found that few intended to change the type of milk consumed [27,35], and most believed the type of milk currently used was best for them [35].

Informal interviews conducted in a New York City Latino community revealed whole milk use represented a symbol of prosperity, a belief derived from the practice in some Latin American countries of diluting milk to stretch resources [36]. A shelf-study in that same community revealed that low-fat milk was available, albeit in smaller quantities than high-fat milk, leading these researchers to conclude that high-fat milk was chosen because it was preferred, not because it was the only choice [37].

In a more recent study, Bus and Worsley documented that Australian shoppers surveyed understood that whole milk had the most fat and calories but mistakenly believed milk was a good source of iron, Vitamin C, and fiber [38]. Many believed "reduced-fat milk" was a good source of vitamins and calcium but more believed this about whole milk. Moreover, shoppers rated whole milk as "tasty" and "feels good in the mouth", and in turn, were more likely to rate reduced-fat milk as "watery" and not "natural".

In summary, over the past decade, the use of low-fat milk has only marginally improved, and there is a large disparity by socio-economic status. Little is known about the knowledge, attitudes and beliefs contributing to this pattern. The literature suggested that the taste and texture of low-fat milk are barriers to its adoption, and, that among some populations, the rich, thicker texture of whole milk may have symbolic meaning. Evaluation of an intervention implemented in the 1990s suggested that promoting the taste, healthiness (less saturated fat), and cost savings of low-fat milk led to increased consumption. On the other hand, a more recent Australian study suggested that although consumers know that low-fat milk contains less saturated fat, they are unaware or are confused about the nutritional content of milk and the differences between types of milk.

If we are to decrease the difference in milk consumption behaviors of low-income Americans so that it is more consistent with the Dietary Guidelines for Americans [2], more research is needed to develop a successful intervention promoting low-fat milk consumption. Many behavior change initiatives fail because the program was developed without a clear understanding of the psychographics of the priority audience or the knowledge, perceptions, beliefs, attitudes, and values of those whose behavior we hope to change. Moreover, in comparison to alternate beverage choices, milk is a nutrient-dense food, and promoting low-fat milk, if it came at the cost of less milk consumption, would not further a healthy eating pattern [10,39]. This would be particularly disconcerting for a low-income population struggling to afford a nutrient-dense diet and avoiding food insecurity [40]. Formative research to understand the perspectives of the priority audience reduces the risk of unintended consequences.

This study had three purposes—to establish the type of milk usually purchased by Oklahoma SNAP recipients (a population with income below 130% of the FPL); to identify sociodemographic characteristics associated with low-fat milk consumption among those SNAP recipients; and to understand SNAP recipients' knowledge, attitudes, and beliefs toward the different types of milk that shape their milk consumption behaviors.

2. Materials and Methods

2.1. Materials

This study is based on a cross-sectional telephone survey of SNAP recipients living in the Oklahoma City media market, which covers 34 of the 77 Oklahoma counties, two of which are considered urban. All procedures were approved by the University of Oklahoma Health Sciences Institutional Review Board. As part of a data sharing agreement, the Oklahoma Department of

Human Services provided identifying information including sociodemographic characteristics of the responsible household member of each SNAP household. This information was routinely collected as part of the SNAP application and semi-annual recertification process. Every respondent gave verbal consent to participate in the study.

2.2. Participants

Among those randomly selected to participate, 553 SNAP recipients were reached, and 520 interviews were completed (94.0% completion rate). Respondents were multi-ethnic and included persons identifying as non-Hispanic White (61.2%), Black (20.8%), Hispanic (8.5%), American Indian (8%), and Asian (1.5%). Most were female (66.6%), single (81.8%), and had attained a high school education or less (81.2%). All study participants had been identified as participating in the SNAP program, which means that all were low-income, having household income of no more than 130% of the FPL or receiving annual income of no more than US $26,668 for a family of four. After weighting the sample proportionally based on ethnicity, one difference was observed between the sample and the Oklahoma SNAP population. Survey respondents were more likely to live in rural areas. However, there were no differences in the types of milk usually purchased between rural and urban areas so this is unlikely to bias the results.

2.3. Measures

Measures were selected from the literature, tested for face and content validity, and pre-tested with people similar to the priority audience. Further, these and other questions were used during the formative focus group research [41], and the ones included in this study had the most probative value. All data were self-reported.

Milk purchasing patterns. Respondents were asked the open-ended question, "Even if you buy or use more than one type of milk, which type of milk do you buy most often for your family"? and were given the opportunity to state what other types of milk were purchased. Responses to both questions were coded as whole, 2%, 1%, nonfat milk, and other (the exact response entered). The survey included three other questions related to milk purchasing patterns, including "How often do you purchase milk?" Open-ended responses were entered as weekly, every two weeks, monthly, and less often. The second question was, "Briefly what is the biggest reason why you usually buy [insert type] milk most often"? (95.3% completion). The third asked whether any changes in the type of milk purchased had been made in the past three months.

Milk nutrition knowledge. A series of 10 true-false questions measured knowledge of general milk nutrition facts. These questions assessed differences in the nutritional attributes of high- and low-fat milk, what types of milk were considered "low-fat", and awareness of the recommended age to serve children low-fat milk.

Stages of change. High-fat milk users' intent to change to low-fat milk was assessed by using stages of change constructs based on the Transtheoretical Model (TTM) [42]. Pre-contemplation was measured with the statements "I would never consider switching to 1% or nonfat milk" and "I have not thought about switching to 1% or nonfat milk". Contemplation was measured with the statement "I am considering switching to 1% or nonfat milk". Preparation was measured using the following two options: "I am ready to switch" and "I sometimes use 1% or nonfat milk".

2.4. Statistical Analysis

Data were analyzed using SPSS Version 20.0. Chi-square test statistics explored bivariate differences in the type of milk usually purchased by sociodemographic variables. A multivariate backward elimination regression model identified sociodemographic characteristics associated with low-fat milk use. A one-way ANOVA compared between-group differences in the mean scores on the milk-nutrition knowledge quiz. The threshold for determining statistical significance was a two-tailed test with 0.05 as the alpha level.

Two coders familiar with the research project independently identified the dominant themes emerging from the open-ended question asking each respondent to explain the reason for purchasing a particular type of milk. Coders began with open coding, assigning themes to ideas and concepts identified in the raw data. These themes were then reviewed for connection and consolidated. While most respondents named only one reason, each theme was coded to ensure that all of the reasons provided were included in the analysis. The themes and concepts that emerged were simple, and there was little coder disagreement.

3. Results

3.1. Milk Consumption Patterns

After excluding those who used other types of milk (1.7%), most SNAP recipients (92.5%) reported usually purchasing high-fat milk (44.6% whole milk and 47.9% purchasing 2% milk) and 7.5% usually purchased low-fat milk (3.9% purchased 1% milk, and 3.6% purchased nonfat milk). In addition, two-thirds of SNAP recipients (67.4%) purchased milk weekly, and 20.0% purchased it every two weeks. Most bought one type of milk (85.3%). Only 4.0% had changed the type of milk usually purchased during the previous three months.

Sociodemographic characteristics were only weakly associated with the type of milk usually purchased among this low-income population. Gender, ethnicity, marital status, residence in a rural or urban area, age, the level of educational attainment, or the presence of a child in the household alone did not significantly differ between high and low-fat milk users. Educational attainment was associated with low-fat milk use if a child lived in the household, $Q_w = 5.6$, df = 1, SE = 0.62, $p = 0.02$. Among SNAP households with children, 14.6% usually purchased low-fat milk if the head of household had attained some college or more compared with 8% where the head of household had a high school education or less. The odds of using low-fat milk were 4.4 times higher (95% CI 1.3, 14.7) in SNAP households with children present when the head of household had attained some college education. There was no association between educational attainment and low-fat milk use if children did not live in the home, $Q_w = 0.12$, df = 1, SE = 0.59, $p = 0.73$.

3.2. Milk Nutrition Knowledge

Overall, test scores on the true-false quiz measuring milk-nutrition knowledge were no better than guessing (M = 51.1, SD = 22.2). On the other hand, low-fat milk users (M = 67.4, SD = 23.0) scored significantly higher on the milk-nutrition quiz than high-fat milk users (M = 49.8, SD = 21.6); $t(494) = 4.7$, $p = 0.00$. Moreover, there was progressive improvement in milk nutrition knowledge among SNAP recipients with each successive reduction in the fat content of the type of milk consumed, $F(1, 492) = 22.9$, $p = 0.00$. Whole milk users had an average score of 46.0 (SD = 21.4), 2% milk users had an average score of 53.2 (SD = 21.3), 1% milk users had an average score of 66.0 (SD = 25.0), and nonfat milk users had an average score of 69.0 (SD = 21.2).

Most SNAP recipients surveyed answered correctly that the fat content differed among the different types of milk (Table 1). Yet, few understood that 2% milk was not low-fat milk. However, 1% milk users were significantly more likely than high-fat milk users to know that 2% milk was not low-fat milk, $X2(1, n = 479) = 6.2$, $p = 0.01$. Other differences in milk nutrition knowledge emerged between high- and low-fat milk users. More SNAP recipients who used low-fat milk understood that 1% milk was not watered-down whole milk, $X2(1, n = 496) = 18.5$, $p = 0.00$. In addition, low-fat milk users had significantly more correct answers than high-fat milk users on questions exploring their general knowledge of the nutrients in the different types of milk—all types of milk have the same amount of calcium, $X2(1, n = 496) = 5.1$, $p = 0.02$, and the same amount of Vitamin D, $X2(1, n = 497) = 7.6$, $p = 0.01$. Similarly, low-fat milk users were more likely to understand that 1% milk has the same vitamins and minerals as whole milk, $X2(1, n = 496) = 6.9$, $p = 0.01$.

Table 1. Correct Answers to Milk Nutrition Knowledge Quiz (weighted).

	Overall (*n* = 496)	High-Fat Milk (Whole and 2%) (*n* = 460)	Low-Fat Milk (1% and Nonfat) (*n* = 36)	*p*-Value
Test Scores	Mean (SE)	Mean (SE)	Mean (SE)	
Overall	51.1 (1.0)	49.8 (3.8)	67.4 (4.0)	0.000 *
Question				
All types of milk have the same amount of calcium.	45.8%	44.3%	63.9%	0.02 *
1% has fewer vitamins and minerals than whole milk.	51.2%	49.6%	72.2%	0.009 *
Whole milk is best for children under two years old.	68.3%	68.5%	66.7%	0.82
1% milk is watered down-whole milk.	43.5%	40.9%	77.8%	0.000 *
Reduced-fat milk and low-fat milk are the same.	50.8%	49.7%	65.7%	0.07
The biggest difference between all types of milk is the percent of fat.	79.6%	79.6%	77.8%	0.79
All children under 18 need whole milk to grow properly.	49.4%	47.3%	77.8%	0.000 *
Whole milk has more Vitamin D than other types of milk.	39.5%	37.7%	61.1%	0.006 *
2% is considered low-fat milk.	23.8%	22.6%	38.9%	0.03 *
1% has fewer calories and less fat but the same vitamins and minerals as whole milk.	59.4%	58.3%	72.2%	0.10

Note: Statistically significant differences in the proportion of correct answers between high and low-fat milk users are denoted with an *.

3.3. Reasons for the Type of Milk Usually Purchased

SNAP respondents were asked to give an open-ended answer that explained their reasoning for choosing the particular type of milk usually purchased. After organizing these responses into themes, the most common reasons related to the perceived health benefits of milk (41.9%) or its taste (30.2%). Some purchased the type of milk based on habit (7.8%). Very few respondents attributed price as an influential factor (1.2%). Responses coded as "other" were vague or not responsive to the question (14.4%). For example, these responses described how milk was used in the household (e.g., "on cereal" or "in cooking") or who usually drank the most milk (e.g., "the older girls drink the 2%, and my infant drinks the whole milk").

However, as illustrated in Table 2, the reason for the type of milk consumed differed by type. Among SNAP recipients who usually purchased whole milk, taste (49.5%) influenced its purchase more often than health benefits (17.7%), and it was the taste and texture of whole milk that was preferred (Table 2). As this respondent explained, "It (whole milk) tastes like milk—the rest tastes like colored water".

Table 2. Reason for Choosing Milk by Type (weighted).

Reason	Whole Milk (*n* = 220)	2% Milk (*n* = 240)	1% Milk (*n* = 20)	Nonfat Milk (*n* = 17)
Taste	49.5%	15.8%	10.0%	11.8%
Health	17.7%	57.1%	85.0%	88.2%
Not concerned with fat	1.4%	0.0%	0.0%	0.0%
Compromise taste/fat	0%	6.3%	0.0%	0.0%
Price	0.5%	2.1%	0.0%	0.0%
Other	22.3%	8.8%	5.0%	0.0%
Habit	8.6%	8.3%	0.0%	0.0%
WIC policy	0.0%	1.7%	0.0%	0.0%

For those SNAP recipients who usually purchased whole milk for health-related reasons, most described whole milk as "healthier" with no further explanation. However, when the health benefits were described, many stated whole milk had more vitamins and minerals than lower-fat milk. For example, explanations included "for the Vitamin D" or "It (has) more Vitamin D in it than the other kinds". Other explanations included "It just feels like I get more vitamins and . . . high protein". Similarly, whole milk was believed to be best for children, regardless of age. As explained by one SNAP recipient, "It (whole milk) has more vitamins for my kids". Other explanations included, "Whole milk is better for the kids because it has more calcium for their bones".

Other common health-related reasons that were reported included the recommendation of a physician or because of a health condition. When whole milk had been recommended by the children's pediatrician, it appeared that this recommendation influenced the type of milk used by the whole family. A few described whole milk as "real", or more "natural".

For 2% milk users, unlike whole milk users, the perceived health benefits (57.1%) influenced the type of milk chosen most often, not taste (Table 2). The health benefits cited most often by these SNAP recipients were less fat and fewer calories, usually less fat. About one in four perceived 2% milk as healthier, stating "it's better for us", "for health", and "it's healthier". Other themes that emerged were that 2% was recommended by a physician or dietician or that 2% was chosen because of a "health condition" such as being diabetic or lactose intolerant.

Two unique reasons emerged among 2% milk users. For a few SNAP recipients, the health benefits of consuming less fat were valued, but the trade-off between taste and health benefits warranted only moving to 2% milk. One respondent exemplified the balance between taste and health benefits, stating "I basically was trying to cut down on fat. Two percent (2%) milk still tastes good; 1% was a little too weak for me". Further, at the time of the survey, WIC policies encouraged low-fat milk use but allowed families with children age 2 and older to use vouchers for the purchase of 2%, 1%, and nonfat milk, which led a few SNAP recipients to choose 2% milk because "that was what WIC allowed".

There was little difference in attitudes between 1% and nonfat milk users. Among those who usually purchased 1% milk, 85.0% said that the biggest reason for this purchase was its health benefits. The frequency increased slightly to 88.2% among nonfat milk users. These SNAP recipients, like many 2% milk users, chose 1% and nonfat milk because they preferred milk with less fat and fewer calories. "Less fat" was expressed as a benefit more often than "fewer calories".

Habit influenced the type of milk consumed among SNAP recipients who used whole or 2% milk, but not those who used low-fat milk. As these whole and 2% milk users explained, "(it was what) I grew up drinking", or "It is just the one I always grab". Deference to maternal figures was also mentioned. For example, one SNAP recipient explained that she used whole milk because that was what her mother had purchased. For other respondents, 2% milk use was a newly formed habit, or at least it was a different type of milk than they had consumed during childhood. For example, one respondent explained, "I just got used to the taste of it. I used to prefer whole milk, but I believe 2% is better for you".

3.4. Stages of Change

Almost three-fourths (74.2%) of high-fat milk users were in the pre-contemplation stages of the Transtheoretical Model of behavior change theory (Figure 1). These SNAP recipients answered that they would never consider switching to 1% or non-fat milk (42.7%), or had not thought about switching to 1% or nonfat milk (31.5%). SNAP recipients who reported using 2% milk were more likely to contemplate or sometimes use low-fat milk than those who used whole milk, $X2(3, n = 453) = 23.9$, $p = 0.00$. Moreover, SNAP recipients with higher scores on the milk nutrition quiz were more likely to be considering or sometimes use low-fat milk, (M = 57.4, SD = 21.6), than those with lower scores, (M = 47.4, SD = 21.1), t(452) = 4.4, $p = 0.00$, 95% CI [14.5, 5.5].

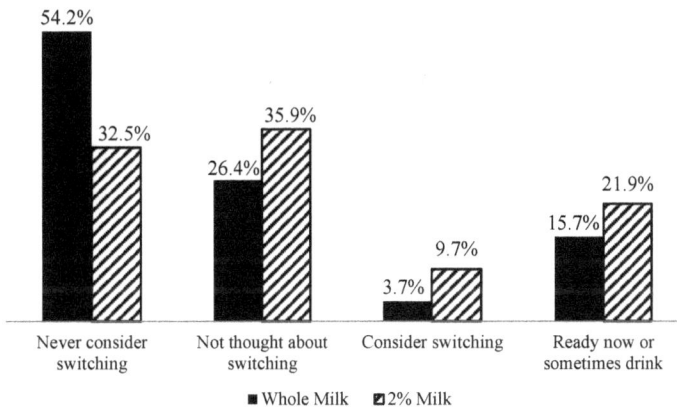

Figure 1. Stages of change constructs (whole and 2% milk users; weighted).

4. Discussion

Based on our findings, low-fat milk use among Oklahoma SNAP recipients was remarkably lower than the 34.4% indicated by national milk sales data [22]. Among this sample of Oklahoma SNAP recipients, 7.5% reported using low-fat milk, a finding comparable to Kitt et al., who found 9.4% of children living in households with income below 130% of the FPL consumed low-fat milk [24]. Moreover, the evidence revealed only a weak association between low-fat milk use and sociodemographic characteristics.

Rather, this study found that the barriers to changing to low-fat milk were associated with poor milk nutrition knowledge and taste preferences. High-fat milk users understood that the fat content differed between the types of milk but did not know that 2% was not low-fat milk; that 1% milk had the same vitamins and minerals of whole milk; and that 1% and nonfat milk were not watered down.

Past interventions, such as the *1% or Less* campaigns of the 1980s and 1990s [28–34], promoted reducing saturated fat as the primary benefit of using low-fat milk. Perhaps, at that time, this nutrition fact was not well-known. However in this study, SNAP recipients usually understood that the main difference between the types of milk was the amount of saturated fat, a finding similar to that of Bus and Worsley [38]. Furthermore, most of the SNAP recipients stated that consuming less saturated fat was the reason they used lower-fat milk (2%, 1% or nonfat milk), suggesting that this was a highly valued benefit, but this alone was not enough for 2% milk users to change to 1% low-fat milk.

Instead, confusion about other milk nutrition facts appeared to be a barrier to adoption of low-fat milk. For example, many believed that 2% milk was low-fat milk, and seemed unaware that choosing 2% milk over whole milk only reduced the saturated fat from 5 g per cup to 3.5 g, a 30% reduction. In comparison, moving from whole to 1% milk reduced the saturated fat to 1.5 g per cup, a 70% reduction. Unfortunately, the nutrition label is either not used or unintelligible to most SNAP recipients. Thus, any promotion must describe the substantial reduction in saturated fat and calories of 1% or nonfat milk, rather than just encouraging the use of "low-fat milk" since this term is not well understood.

Equally, if not more importantly, a barrier to the adoption of low-fat milk was the misperception that low-fat milk had fewer vitamins and minerals than high-fat milk. The reasons given for the usual type of milk purchased revealed low-fat milk was perceived as watery—a diluted milk with fewer of the valued vitamins and minerals of milk, such as calcium and Vitamin D. Answers to specific questions on the milk nutrition quiz reinforced this finding.

Product familiarity is another known motivation for food choice among low-income populations [43] and our findings indicate that use of higher-fat milks, particularly whole milk,

Beverages **2017**, *3*, 46

was an entrenched habit among the Oklahoma SNAP population. Many SNAP beneficiaries used the same type of milk as consumed during childhood, and few had changed the type of milk used during the three-month period preceding the study. Similar to Tuorila [35], most high-fat milk users were not contemplating changing to lower-fat milk.

Regardless, 2% milk users showed more willingness to consider changing to low-fat milk than whole milk users. They valued the benefit of consuming less saturated fat. All of these facts suggest that current 2% milk users will be more receptive to efforts to promote 1% milk consumption.

Previous studies have concluded that price is an influential motive for choosing foods among low-income populations [43,44]. Cost savings were also used successfully to promote low-fat milk in the *1% or Less* campaigns [30,31]. However, low-fat milk usually costs less in Oklahoma than higher fat milk and few SNAP recipients sought out these price savings. Ariely posits that the quality of a product is subjectively discounted when it costs less than similar items [45]. This study documents that most SNAP recipients believed low-fat milk is inferior to higher-fat milk. It is unclear whether the lower price predictably but irrationally influenced this negative evaluation or whether the price differential is too small to overcome the negative evaluation of low-fat milk. More research is needed to understand why these SNAP recipients were not motivated enough by price to choose 1% or nonfat milk.

5. Conclusions

People with low-incomes in the US are not considered to be an important consumer group, and marketing research ignores them on a variety of consumer issues. As a consequence, we know little about what motivates their consumer behavior. This is a crucial deficit for any effort to improve the diet of America's low-income population and reduce food insecurity by maximizing the nutritional value of the foods consumed consistent with the Dietary Guidelines for Americans [2]. As this study has documented, low-fat milk use among Oklahoma SNAP recipients lags behind national trends. However, with the exception of educational attainment when children are present in the household, sociodemographic characteristics were not a significant influence on the type of milk consumed. Instead, less low-fat milk consumption was associated with a knowledge gap and misperceptions of the nutritional properties of different types of milk.

For the purposes of promoting low-fat milk use, messages should emphasize three key facts that differentiate high- and low-fat milk consumers: (1) 2% milk is not low-fat milk; (2) 1% milk is not watered down; and (3) 1% low-fat milk has all the vitamins and minerals of whole milk. These messages should resonate better with 2% milk users because they have already adopted what they consider a healthier behavior and are more open to change than whole milk users who are more resolute. Finally, what is considered low-fat milk is not well understood by SNAP recipients. Therefore, the emphasis should be placed on consuming 1% milk, and not on the confusing general category of low-fat milk.

Acknowledgments: Partial support for this study came from the US Department of Agriculture through the Oklahoma Department of Human Services through a Supplemental Nutrition Assistance Program, Nutrition Education Grant (Robert John, PI). This research was a component of a 1% low-fat milk intervention that received the National Social Marketing Centre (Great Britain) Award for Excellence in Social Marketing announced at the University of South Florida's 24th Social Marketing Conference, 2016. The funding agencies had no role in the design, analysis, or writing of this article. The interpretations expressed here are the authors and do not reflect the views of the funding agencies.

Author Contributions: Both authors listed contributed substantially to this project and manuscript. R.J. and K.J.F. conceived and designed the study; K.J.F. assembled the data and performed the data analysis; R.J. and K.J.F. made substantial contributions to writing the manuscript; and both authors reviewed the manuscript.

Conflicts of Interest: The authors declare no conflict of interest. The funding agencies had no role in the design of the study, in the collection, analysis, or interpretation of data, or the writing of the manuscript.

References

1. Colemen-Jensen, A.; Rabbitt, M.P.; Gregory, C.A.; Singh, A. *Household Food Security in the United States in 2015*; ERS Rep No. 215; Department of Agriculture: Washington, DC, USA, 2016.
2. US Department of Health and Human Services and US Department of Agriculture. *2015–2020 Dietary Guidelines for Americans*, 8th ed.; Human Nutrition Information Service: Washington, DC, USA, 2015. Available online: http://health.gov/dietaryguidelines/2015/guidelines/ (accessed on 1 February 2016).
3. Drewnowski, A. The nutrient rich foods index helps to identify healthy, affordable foods. *Am. J. Clin. Nutr.* **2010**, *91S*, 1095S–1101S.
4. Drewnowski, A. The contribution of milk and milk products to micronutrient density and affordability of the US diet. *J. Am. Coll. Nutr.* **2011**, *30*, 422S–428S. [PubMed]
5. Drewnowski, A.; Specter, S.E. Poverty and obesity: The role of energy density and energy costs. *Am. J. Clin. Nutr.* **2004**, *79*, 6–16. [PubMed]
6. Huth, P.J.; Fulgoni, V.L.; Keast, D.R.; Park, K.; Auestad, N. Major food sources of calories, added sugars, and saturated fat and their contribution to essential nutrient intakes in the US diet: Data from the National Health and Nutrition Examination Survey (2003–2006). *Nutr. J.* **2013**, *12*, 1–10.
7. Huth, P.J.; Fulgoni, V.L., III; DiRienzo, D.B.; Miller, G.D. Role of dairy foods in the dietary guidelines. *Nutr. Today* **2008**, *43*, 226–234.
8. Sebastian, R.S.; Goldman, J.D.; Enns, C.W.; LaComb, R. *Fluid Milk Consumption in the United States: What We Eat in America, NHANES 2005–2006*; FSRG Dietary Data Brf No. 3; Agriculture Research Service: Beltsville, MD, USA, 2010. Available online: http://ars.usda.gov/Services/docs.htm?docid=19476 (accessed on 10 November 2010).
9. Aggarwal, A.; Monsivais, P.; Drewnowski, A. Nutrient intakes linked to better health outcomes are associated with higher diet costs in the US. *PLos ONE* **2012**, *7*, e37533.
10. Rehm, C.D.; Drewnowski, A.; Monsivais, P. Potential population-level nutritional impact of replacing whole and reduced-fat milk with low-fat and skim milk among US children aged 2–19 years. *J. Nutr. Educ. Behav.* **2015**, *47*, 61–68. [PubMed]
11. Mancino, L.; Guthrie, J. *SNAP Households Must Balance Multiple Priorities to Achieve a Healthful Diet*; Amber Waves; Economic Research Service: Washington, DC, USA, 2014. Available online: http://www.ers.usda.gov/amber-waves/2014-november/snap-households-must-balance-multiple-priorities-to-achieve-a-healthful-diet.aspx#.VYrqhflVhBc (accessed on 24 June 2015).
12. Huth, P.J.; Park, P.M. Influence of dairy product and milk fat consumption on cardiovascular disease risk: A review of the evidence. *Adv. Nutr.* **2012**, *3*, 266–285. [PubMed]
13. Kratz, M.; Baars, T.; Guyenet, S. The relationship between high-fat dairy consumption and obesity, cardiovascular, and metabolic disease. *Eur. J. Nutr.* **2013**, *52*, 1–24. [PubMed]
14. Hammad, S.; Pu, S.; Jones, P.J. Current evidence supporting the link between dietary fatty acids and cardiovascular disease. *Lipids* **2016**, *51*, 507–517. [PubMed]
15. Stamler, J. Diet-heart: A problematic revisit. *Am. J. Clin. Nutr.* **2010**, *91*, 497–499. [PubMed]
16. O'Sullivan, T.A.; Hafekost, K.; Mitrou, F.; Lawrence, D. Food sources of saturated fat and the association with mortality: A meta-analysis. *Am. J. Public Health* **2013**, *103*, e31–e42. [PubMed]
17. Drehmer, M.; Pereira, M.A.; Schmidt, M.I.; Alvim, S.; Lotufo, P.A.; Luft, V.C.; Duncan, B.B. Total and full-fat, but not low-fat, dairy product intakes are inversely associated with metabolic syndrome in adults. *J. Nutr.* **2016**, *146*, 81–89. [PubMed]
18. US Department of Agriculture and the US Department of Health and Human Services. *Scientific Report of the 2015 Dietary Guidelines Advisory Committee*; Human Nutrition Information Service: Washington, DC, USA, 2015. Available online: http://www.health.gov/dietaryguidelines/2015-scientific-report/PDFs/Scientific-Report-of-the-2015-Dietary-Guidelines-Advisory-Committee.pdf (accessed on 1 February 2016).
19. US Department of Agriculture Services and US Department of Health and Human Services. *Dietary Guidelines for Americans 2010*, 7th ed.; Human Nutrition Information Service: Washington, DC, USA, 2011. Available online: https://health.gov/dietaryguidelines/dga2010/dietaryguidelines2010.pdf (accessed on 20 November 2010).

20. US Department of Agriculture Services and US Department of Health and Human Services. *Dietary Guidelines for Americans, 2005*, 6th ed.; Human Nutrition Information Service: Washington, DC, USA, 2005. Available online: https://health.gov/dietaryguidelines/dga2005/document/ (accessed on 12 April 2017).

21. Britten, P.; Marco, K.; Juan, W.; Guenther, P.M.; Carlson, A. *Trends in Consumer Food Choices within the MyPyramid Milk Group*; Nutr Insights 35; USDepartment of Agriculture and Center for Nutrition Policy and Promotion: Alexandria, VA, USA, 2007. Available online: http://www.cnpp.usda.gov/Publications/NutritionInsights/Insight35.pdf (accessed on 10 November 2011).

22. US Department of Agriculture. *Fluid Milk Sales by Product (Annual)*; Economic Research Service: Washington, DC, USA, 2013. Available online: http://www.ers.usda.gov/data-products/dairy-data.aspx (accessed on 5 May 2017).

23. Davis, C.G.; Dong, D.; Blayney, D.; Yen, S.T.; Stillman, R. US fluid milk demand: A disaggregated approach. *Int. Food Agribus. Manag. Rev.* **2012**, *15*, 25–40.

24. Kitt, B.K.; Carroll, M.D.; Ogden, C.L. *Low-Fat Milk Consumption among Children and Adolescents in the United States, 2007–2008*; NCHS Data Brief; National Center for Health Statistics: Hyattsville, MD, USA, 2011; pp. 1–8.

25. Maglione, C.; Barnett, J.; Maddock, J.E. Correlates of low-fat milk consumption in a multi-ethnic population. *Calif. J. Health Promot.* **2005**, *3*, 21–27.

26. Caster, L.; Mabli, J. *Food Expenditures and Diet Quality among Low-Income Households and Individuals*; Mathematica Policy Research, Inc.: Washington, DC, USA, 2010.

27. Brewer, J.L.; Blake, A.J.; Rankin, S.A.; Douglass, L.W. Theory of reasoned action predicts milk consumption in women. *J. Am. Diet Assoc.* **1999**, *99*, 39–44. [PubMed]

28. Booth-Butterfield, S.; Reger, B. The message changes belief and the rest is theory: The *1% or Less* milk campaign and reasoned action. *Prev. Med.* **2004**, *39*, 581–588. [PubMed]

29. Reger, B.; Wootan, M.; Booth-Butterfield, S. Using mass media to promote healthy eating: A community-based demonstration project. *Prev. Med.* **1999**, *29*, 414–421. [PubMed]

30. Reger, B.; Wootan, M.; Booth-Butterfield, S. A comparison of different approaches to promote community-wide dietary change. *Am. J. Prev. Med.* **2000**, *18*, 271–275. [PubMed]

31. Reger, B.; Wootan, M.G.; Booth-Butterfield, S.; Smith, H. *1% or Less*: A community-based nutrition campaign. *Public Health Rep.* **1998**, *113*, 410–419. [PubMed]

32. Reger-Nash, B.; Wootan, M.G.; Booth-Butterfield, S.; Cooper, L. The cost-effectiveness of *1% or Less* media campaigns promoting low-fat milk consumption. *Prev. Chronic Dis.* **2005**, *2*, A05.

33. Hinkle, A.J.; Mistry, R.; McCarthy, W.J.; Yancey, A.K. Adapting the *1% or Less* milk campaign for a Hispanic/Latino population: The adelante con leche semi-descremada 1% experience. *Am. J. Health Promot.* **2008**, *23*, 108–111. [PubMed]

34. Maddock, J.E.; Maglione, C.; Barnett, J.D.; Cabot, C.; Jackson, S.; Reger-Nash, B. Statewide implementation of the *1% or Less* campaign. *Health Educ. Behav.* **2007**, *34*, 953–963. [PubMed]

35. Tuorila, H. Selection of milks with varying fat contents and related overall liking, attitudes, norms and intentions. *Appetite* **1987**, *8*, 1–14. [PubMed]

36. Wechsler, H.; Wernick, S.M. A social marketing campaign to promote low-fat milk consumption in an inner-city Latino community. *Public Health Rep.* **1992**, *107*, 202–207. [PubMed]

37. Wechsler, H.; Basch, C.E.; Zybert, P.; Shea, S. The availability of low-fat milk in an inner-city Latino community: Implications for nutrition education. *Am. J. Public Health* **1995**, *85*, 1690–1692. [PubMed]

38. Bus, A.E.; Worsley, A. Consumers' sensory and nutritional perceptions of three types of milk. *Public Health Nutr.* **2002**, *6*, 201–208.

39. Nolan-Clark, D.; Mathers, E.; Probst, Y.; Charlton, K.; Tapsell, L.C. Dietary consequences of recommending reduced-fat dairy products in the weight-loss context: A secondary analysis with practical implications for registered dietitians. *J. Acad. Nutr. Diet.* **2012**, *113*, 452–458.

40. Drewnowski, A.; Darmon, N. The economics of obesity: Dietary energy density and energy cost. *Am. J. Clin. Nutr.* **2005**, *82*, 265S–273S. [PubMed]

41. Finnell, K.J.; John, R. Formative research to understand the psychographics of 1% milk consumption in a low-income audience. *Soc. Mark. Q.* **2017**, *23*, 169–184.

Beverages **2017**, 3, 46

42. Prochaska, J.O.; Redding, C.A.; Evers, K.E. The transtheoretical model and stages of change. In *Health Behavior: Theory, Research, Practice*, 5th ed.; Glanz, K., Rimer, B.K., Viswanath, K., Eds.; John Wiley & Sons, Inc.: San Francisco, CA, USA, 2015.

43. Steptoe, A.; Pollard, T.M.; Wardle, J. Development of a measure of the motives underlying the selection of food: The Food Choice Questionnaire. *Appetite* **1995**, 25, 267–284. [PubMed]

44. John, R.; Kerby, D.S.; Landers, P.S. A market segmentation approach to nutrition education among low-income individuals. *Soc. Mark. Q.* **2004**, 10, 24–38.

45. Ariely, D. *Predictably Irrational: The Hidden Forces That Shape Our Decisions*; Harper Perennial: New York, NY, USA, 2009.

beverages

MDPI

Short Note

Bioactive Peptides in Milk: From Encrypted Sequences to Nutraceutical Aspects

Massimo Lucarini

Consiglio per la Ricerca in Agricoltura e l'Analisi dell'Economia Agraria—Centro di Ricerca Alimenti e Nutrizione (CREA-AN), Via Ardeatina 546, 00178 Roma, Italy; massimo.lucarini@crea.gov.it; Tel.: +39-065-149-4446

Academic Editor: Alessandra Durazzo
Received: 30 May 2017; Accepted: 2 August 2017; Published: 14 August 2017

Abstract: Milk provides a wide range of biologically active compounds that protect humans against diseases and pathogens. The purpose of this work is to describe the main aspects and research lines concerning bioactive peptides: from their chemistry, bioavailability, and biochemical properties to their applications in the healthcare sector. In this context, the uses of bioactive peptides in nutraceutical and functional foods have been highlighted, also taking into account the perspective of innovative applications in the field of circular bioeconomy.

Keywords: bioactive peptides; milk; nutraceuticals; bioavailability; biorefinery

1. Bioactive Components of Milk: Focus on BPs

Milk provides a wide range of biologically active compounds that protect humans against diseases and pathogens such as immunoglobulins, antimicrobial proteins and peptides, oligosaccharides, lipids, as well as many other components at low concentrations with significant potential health benefits [1] (Figure 1). Among these compounds, proteins play a key role as nutrients and as promoters of the physiological function and health role as source of Bioactive Peptides (BPs) [2].

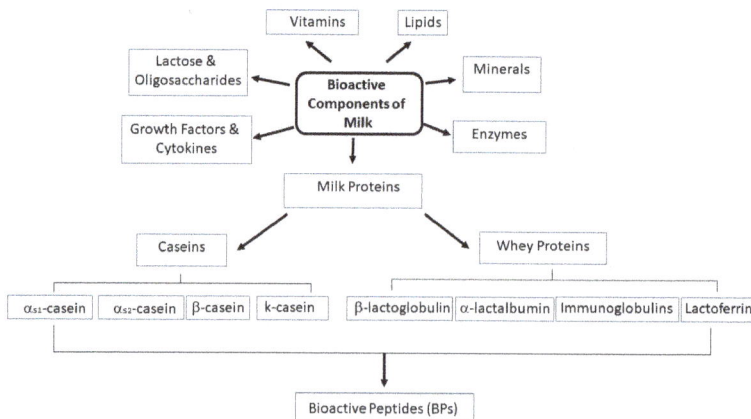

Figure 1. Major bioactive functional compounds derived from milk (adapted fom [1]).

In cow's milk, around 80% of the protein present is casein (CN) and 20% is whey protein. CN consists of αs1-, αs2-, β- and k-CN families in the approximate ratio 38:11:38:13 [3]. Whey proteins are typically a mixture of beta-lactoglobulin (~65%), alpha-lactalbumin (~25%), bovine serum albumin (~8%), lactoferrin, and immunoglobulins. The value of proteins as an essential source of amino acids has been well documented, but in the 1980s, it was recognized that dietary proteins can exert many other in vivo functions through BPs. The isolation of endogenous opioid peptides called enkephalins, which occurred for the first time in 1975, led to the discovery of the opioid peptide activity derived from partial enzymatic digestion of milk proteins [4]. Since then, the bioactive components derived from food proteins, in particular from milk proteins, have been subjected to numerous studies focused on their structural and biochemical properties.

2. Biochemical Properties

BPs peptides are encrypted and inactive in the parent protein sequence but can be released and activated through the enzymatic proteolysis (gastrointestinal digestion, in vitro hydrolysis using proteolytic enzymes) of proteins and during food processing (cooking, fermentation, ripening). Once released, the BPs can act as regulatory compounds such as β-casomorphins and casein-derived phosphopeptides and are released in vivo during the gastrointestinal transit of milk and dairy products [5]. It is important to underline the strong connection between the structure and the biological activities: the amino acid sequence, the hydrophobicity, the charge of BPs determine a specific activity.

For instance, in the case of β-casomorphins (BCMs), the presence of a tyrosine residue at the amino terminal end (except δ-casein opioids) and the presence of aromatic amino acids in the third and fourth positions of the peptide represents a structural feature of the opioid activity; in particular, the phenolic hydroxyl group of tyrosine gives rice a negative potential, essential for the opioid activity. On the other hand, the Pro residue maintains the proper orientation of the Tyrosine and Phenylalanine side chains [6,7]. In regard to the importance of the amino acid sequence on the activity of BPs, human BCMs (Tyr-Pro-Phe-Val-Glu-Pro-Ile) were found to be 3 to 30 times less potent than bovine BCMs (Tyr-Pro-Phe-Pro-Gly-Pro-Ile) [7].

The production and properties of BPs have been reviewed in many articles [8–10]. After ingestion, BPs can affect the cardiovascular, nervous, gastrointestinal, and immunological systems. BPs have been defined as specific protein fragments having a positive influence on the physiological and metabolic functions or conditions of the body. Therefore, they may have crucial beneficial effects on human health [11].

Furthermore, it is worth mentioning the recent review of Nielsen et al. [12], where a comprehensive database of milk protein–derived BPs that can be used to search for specific functions—those of peptides or proteins—was developed. A new visual arrangement was proposed: BPs were visually mapped on the basis of parent protein sequences, providing information on sites with the highest abundance of these compounds (http://mbpdb.nws.oregonstate.edu/).

Their main regulatory effects concern the following:

— the transport of minerals (caseinophosphopeptides), such as calcium, and intestinal transport of amino acids, such as leucine, through the beta-casomorphin receptors;
— the transport of intestinal fluid (beta-casomorphine);
— the motility of the gastrointestinal tract (beta-casomorphine);
— the stimulation of the postprandial hormone secretion (insulin, somatostatin) (beta-casomorphine);
— the regulation of insulin secretion based on glucose concentration;
— immunostimulant peptides (alpha and beta casein fragments);
— anti-hypertensive peptides enzyme inhibitors converting angiotensin I (ACE) (casokinine);
— antithrombotic peptides such as ADP-activated platelet aggregation inhibitors, as well as fibrinogen binding (γ-chain) to ADP-treated platelets (casoplateline);
— opioid activities;
— antioxidative functions;
— hypocholesterolemic activities;
— antitumor activities.

Table 1. Examples of bioactive peptides from milk.

Precursor Protein	Fragment	Peptide Sequence	Name	Biological Activity	Preparation	References
Casein Protein						
β-casein	60–70	YPFPGPIPNSL	β-casomorphin-11	Opioid	Hydrolysis with digestive enzymes Trypsin	[13]
	60–66	YPFPGPI	β-casomorphin-7	Opioid ACE Inhibitory Immunomodulatory	Mixture of gastro-intestinal enzymes Trypsin	[14]
	60–64	YPFPG	β-casomorphin-5	Opioid ACE Inhibitory	Hydrolysis with trypsin	[15]
	177–183	AVPYPQR	β-casokinin-7	ACE Inhibitory	Hydrolysis with trypsin	[15]
	193–202	YQQPVLGPVR	β-casokinin-10	ACE Inhibitory Immunomodulatory	Hydrolysis with trypsin	[16]
	169–175	KVLPVPQ		ACE inhibition	Hydrolysis with proteinase	[17]
	63–68	PGPIPN	Immunopeptide	Immunomodulatory	Trypsin or chymosin	[18]
	191–193	LLY	Immunopeptide	Immunomodulatory	Trypsin or chymosin	[18]
	114–118	YPVEP	βcasochemotide-1	Immunomodulatory	Hydrolysis with proteinase	[19]
	210–221	EPVLGPVRGPFP		ACE-inhibition	Fermentation	[20]
αs_1-casein	(1-25)4P	RELEELNVPGEIVESLSSSEESITR	Caseinophosphopeptide	Ca++ binding	Trypsin or chymosin	[21]
	90–96	RYLGYLE	α-casein exorphin	Opioid	Hydrolysis with pepsin	[22]
	90–95	RYLGYL	α-casein exorphin	Opioid	Hydrolysis with pepsin	[22]
	91–96	YLGYLE	α-casein exorphin	Opioid	Hydrolysis with pepsin	[22]
	23–27	FFWAP	αs_1-Casokinin-5	ACE inhibition	Hydrolysis with trypsin	[16]
	28–34	FPEWFGK	αs_1-Casokinin-7	ACE inhibition	Hydrolysis with trypsin	[16]
	194–199	TTMPLW	αs_1-Casokinin-6	ACE inhibition, Immunomodulatory	Hydrolysis with trypsin	[23]
	169–193	LGTQYTDAPSFSDIPNPIGSENSEK		ACE-inhibition	Trypsin or chymosin	[24]

Table 1. *Cont.*

Precursor Protein	Fragment	Peptide Sequence	Name	Biological Activity	Preparation	References
αs2-casein	94–103	QKALNEINQF		Antimicrobial ACE inhibition	Hydrolysis with chymotrypsin	[25]
	163–176	TKKTKLTEEKNRL		ACE inhibition	Hydrolysis with chymotrypsin	[25]
	33–38	SRYPSY	Casoxin 6	Anti-Opioid	Hydrolysis with pepsin	[26]
	25–34	YIPIQYVLSR	Casoxin C	Anti-Opioid	Hydrolysis with trypsin	[27]
k-Casein	106–116	MAIPPKKNQDK	Casoplatelin	Antithrombotic:inhibition of platelet aggragation	Hydrolysis with trypsin	[28]
		YIPSY	Casoxin 4	Opioid agonist	Synthetic	[29]
Whey Proteins						
α-lactalbumin	50–53	YGLF	α-lactorphin	Opioid agonist ACE inhibition	Hydrolysis with gastric and pancreatic enzymes	[30]
β-lactoglobulin	102–105	TLLF	β-lactorphin	Non-opioid ACE-inhibition	Tryptic digest	[31]
	142–148	ALPMHIR		ACE-inhibition	Proteolytic digestion	[32,33]
β-lactoglobulin	146–149	HIRL	β-lactotensin	Ileum contraction, hypocholesterolemic activity	Synthetic	[31,34]
Bovine Serum Albumin	208–216	ALKAWSVAR	Albutensin A	Ileum contraction, ACE inhibition	Hydrolysis with proteinase	[32]
	399–404	YGFQDA	Serorphin	Opioid	Hydrolysis with pepsin	[35]
Lactoferrin	17–41/42	FKCRRWQWRMKKLGAPSICURRAF/A	Lactoferricin	Antimicrobial	Hydrolysis with pepsin	[36]

Recent studies have suggested that milk BPs may also contribute to reducing the risk of obesity and development of metabolic disorders. The size of these peptides may vary from two to more than 20 amino acids residues, and each defined peptide bioactivity is strictly linked to its structural features. In Table 1, some examples of the main BPs derived from casein and whey milk proteins are reported.

Some milk BPs show multifunctional properties [37]; for example, some regions in the primary structure of caseins contain overlapping peptide sequences that exert different biological effects (Table 1). These regions have been considered as 'strategic zones' which are partially protected from the proteolytic breakdown [38,39]; e.g., peptides from the sequence 60–70 (β-casomorphin-11) and 60–66 (β-casomorphin-7) of β-casein show immunostimulatory, opioid, and ACE-inhibitory activities. These sequences are protected from proteolysis because of their high hydrophobicity, which is due to the presence of proline and other hydrophilic amino acids. Neutral and basic amino acids are, instead, rapidly hydrolyzed. Moreover, it is interesting to highlight that some bioactive sequences include proline in their domain.

3. Bioavailability

The amount of peptides released upon digestion, as well as the beneficial effects of human health, are hardly predictable. It was estimated that the theoretical yield of opioid peptides encrypted in milk proteins ranged between 2% (β-Casomorphin-5 from β-Casein, f60–64) and 6% (α-Lactorphin from α-Lactalbumin, f50–53) starting from the precursor peptide [40]. BPs bioavailability is the ability of peptides to exert physiological effects in vivo after oral ingestion. Thus, it is of crucial importance that milk-derived BPs remain active during the gastrointestinal digestion and absorption and reach the target site intact. This means that milk-derived BPs have to be resistant to hydrolysis in the gastrointestinal tract in order to reach the peripheral organs [41]. The bioavailability of peptides depends on a variety of structural and chemical properties, i.e., resistance to proteases, charge, molecular weight, hydrogen bonding potential, hydrophobicity, and the presence of specific residues [42–44]. Indeed, proline- and hydroxyproline-containing peptides are relatively resistant to degradation by digestive enzymes [45–47]. The composition of the intestinal content, including food, significantly varies. The time that a peptide is present in the GI tract, as well as its absorption, are significantly affected by gastric emptying and intestinal transit. In addition, the peptide transport could be inhibited or favored (related to pKa of the peptide) by the physiological pH.

For example, some milk-derived peptides have an in vitro inhibitory ACE activity compared to the relative synthetic ACE inhibitor, but exhibit high in vivo activity. This behavior was attributed to a greater affinity of BPs for tissues and their slower elimination [48], but other modes of action were hypothesized [16]. In contrast, some ACE-inhibitory peptides exhibit high activity in vitro but have no effects in vivo. For example, the peptide FFWAP derived from αs1-CN [49–51] is a potent ACE inhibitor in vitro but has no hypotensive effect in vivo [15]. Generally, the difficulty of establishing a direct relationship between in vitro and in vivo activity may depend on different reasons but it is clear that bioavailability after oral administration plays a key role. Despite numerous "in vitro" studies, further research is needed to clarify the relationship between the activity of BPs and their bioavailability. In this regard, the recent review by Nongonierma and FitzGerald, [52] summarizes the scientific evidence for the role of milk protein-derived BPs in humans and pointed out how double-blind randomized clinical trials based on the use of universal guidelines for the evaluation of BPs in humans are needed.

4. Nutraceutical Aspects

Numerous bioactive peptide fragments can be obtained through the hydrolysis of whole milk or via the precursor protein by digestive enzymes. This strongly leads us to hypothesize the production of such peptides in the GIT after the consumption of milk.

For these reasons, the potential that different BPs have to improve human health by reducing the risk of chronic diseases or by activating the immune response is gaining more and more interest within the scientific and commercial fields. In fact, since milk proteins are a rich source of natural active compounds, they could be used as ingredients for different applications and represent nutraceutical substances with potential health benefits that make them suitable for food and pharmaceutical applications [53].

4.1. Methodology toward Innovative Aspects

Processing techniques in laboratory and on industrial scale are now being developed for the extraction, fractionation and isolation of main proteins from milk [54–57], with particular attention to modern non-thermal, clean and green methodologies [58]. It is worth stressing, as already demonstrated by several studies, that each time a nutraceutical with potential health benefits is needed for a functional food, a specific enzymatic hydrolysis of milk proteins should be planned [59,60]. It is important to mention the review by Hafeez et al. [61] on different strategies for increasing the production of BPs from milk in order to obtain functionalized fermented products; the authors reported three types of methodology approaches: (1) proteolytic system of lactic acid bacteria (LAB) or food grade enzymes or a combined dual approach, (2) supplementation of the fermented milk products with the BPs obtained outside of the product, and (3) microorganisms using recombinant DNA technology [61]. Linares et al. [62] reviewed and discussed the use of health-supporting bacteria as starters or adjunct cultures for the development of dairy foods with targeted functional properties, including BPs.

4.2. Applications

The recent review by Mohanty et al. [63] gives a preliminary classification of bioactive milk-derived peptides and their impact on human health. It describes their physiological functions, general characteristics, and potentialities for improving health, as well as their nutraceutical and/or pharmaceutical applications. Besides their biochemical and physiological efficacy and versatility, milk-derived BPs are considered as ingredients of functional foods, as pointed out in a review by Park and Nam [64]. Milk BPs in dairy and non-dairy food formulations have been exploited, as shown by several authors [65–67]. The fractionation of bioactive milk ingredients represents a new, emerging sector of the market, which offers new and innovative products [53,59,67]. FitzGerald, et al. [68] exploited the hypotensive peptides from milk proteins and reported some examples of hypotensive dairy protein-derived products on the market such as a sour milk named "Calpis" (Calpis Co., Tokyo, Japan) or a fermented milk called "Evolus®" (Valio Oy, Helsinki, Finland) containing IPP and VPP peptides (peptides with hypotensive effect) from β- and κ-casein. CalpisTM and Evolus® have been tested extensively in rats and in a clinical trial [69].

In addition, Korhonen, and Pihlanto, [70] have reported other commercial dairy products and ingredients having potential health benefits due to the presence of BPs, such as "BioZate" (Davisco Food International, Eden Prairie, MN, USA). It is a hydrolyzed whey protein that claims to reducing blood pressure. Another example is represented by "Vivinal Alpha," an ingredient/hydrolysate (Borculo Domo Ingredients—BDI-, Zwolle, The Netherlands) that favors relaxation and sleep.

Today, the recovery of BPs from milk and dairy industry byproducts has recently caught the attention of researchers in industrial biorefineries [71–73]. In order to solve the problem of the high environmental impact of byproducts deriving from cheese production industries, several new techniques are now employed to convert milk and dairy waste and byproducts into value-added products.

It is worth mentioning Patel et al. [74], who underline the emerging trends in nutraceutical applications of whey protein and its derivatives. To cite an example, Athira et al. [75] produced and characterized a whey protein hydrolysate with an antioxidant activity from cheese whey using the response surface methodology. In recent research, Abd El-Salam and El-Shibiny [76] have summarized today's ongoing research into the preparation, properties, and uses of enzymatic milk protein hydrolysates.

5. Conclusions

The exploitation of the chemistry, bioavailability, and biochemical properties of BPs in cow's milk represent the basis for the development of nutraceutical and functional foods, in particular in the perspective of innovative applications in the field of circular bioeconomy. Moreover, for the use of these BPs as nutraceuticals, it is important to encourage clinical trials to test their effectiveness on humans and support the research through projects funded by public and private institutions.

Acknowledgments: The author thanks Annalisa Lista for the linguistic revision and editing of this paper.

Conflicts of Interest: The authors declare no conflict of interest.

References

1. Park, Y.W. Overview of bioactive components in milk and dairy products. In *Bioactive Components in Milk and Dairy Products*; Park, Y.W., Ed.; Wiley-Blackwell Publishers: Oxford, UK, 2009; pp. 3–14.
2. Baum, F.; Fedorova, M.; Ebner, J.; Hoffmann, R.; Pischetsrieder, M. Analysis of the endogenous peptide profile of milk: Identification of 248 mainly casein-derived peptides. *J. Proteome Res.* **2013**, *12*, 5447–5462. [CrossRef] [PubMed]
3. Walstra, P.; Jenness, R. *Dairy Chemistry and Physics*; John Wiley: New York, NY, USA, 1984.
4. Teschemacher, H.; Csontos, K.; Westenthanner, A.; Brantl, V.; Kromer, W. Endogenous opioids: Cold-induced release from pituitary tissue in vitro; extraction from pituitary and milk. In *Endorphins in Mental Health Research*; Usdin, E., Bunney, W.E., Kline, N.S., Eds.; Palgrave Macmillan: Basingstoke, UK, 1979; pp. 203–208.
5. Phelan, M.; Aherne, A.; FitzGerald, R.J.; O'Brien, N.M. Casein-derived peptides: Biological effects, industrial uses, safety aspects and regulatory status. *Int. Dairy J.* **2009**, *19*, 643–654. [CrossRef]
6. Kamiński, S.; Cieslińska, A.; Kostyra, E. Polymorphism of bovine beta-casein and its potential effect on human health. *J. Appl. Genet.* **2007**, *48*, 189–198. [CrossRef] [PubMed]
7. European Food Safety Authority (EFSA). Review of the potential health impact of β-casomorphins and related peptides. *Eur. Food Saf. Auth.* **2009**, *7*. [CrossRef]
8. Lopez-Fandino, R.; Otte, J.; van Camp, J. Physiological, chemical and technological aspects of milk-protein-derived peptides with antihypertensive and ACE inhibitory activity. *Int. Dairy J.* **2006**, *16*, 1277–1293. [CrossRef]
9. Kamau, S.M.; Lu, R.-R.; Chen, W.; Liu, X.-M.; Tian, F.-W.; Shen, Y.; Gao, T. Functional significance of bioactive peptides derived from milk proteins. *Food Rev. Int.* **2010**, *26*, 386–401. [CrossRef]
10. Nagpal, R.; Behare, P.; Rana, R.; Kumar, A.; Kumar, M.; Arora, S.; Morotta, F.; Jain, S.; Yadav, H. Bioactive peptides derived from milk proteins and their health beneficial potentials: An update. *Food Funct.* **2011**, *2*, 18–27. [CrossRef] [PubMed]
11. Kitts, D.D.; Weiler, K. Bioactive proteins and peptides from food sources. Applications of bioprocesses used in isolation and recovery. *Curr. Pharm. Des.* **2003**, *9*, 1309–1323. [CrossRef] [PubMed]
12. Nielsen, S.D.; Beverly, R.L.; Qu, Y.; Dallas, D.C. Milk bioactive peptide database: A comprehensive database of milk protein-derived bioactive peptides and novel visualization. *Food Chem.* **2017**. [CrossRef] [PubMed]
13. Meisel, H. Chemical characterization and opioid activity of an exorphin isolated from in vivo digests of casein. *FEBS Lett.* **1986**, *196*, 223–227. [CrossRef]
14. Cieslińska, A.; Kostyra, E.; Kostyra, H.; Oleński, K.; Fiedorowicz, E.; Kamiński, S. Milk from cows of different β-casein genotypes as a source of β-casomorphin-7. *Int. J. Food Sci. Nutr.* **2012**, *63*, 426–430. [CrossRef] [PubMed]
15. Maruyama, S.; Nakagomi, K.; Tomizuka, N.; Suzuki, H. Angiotensin I-converting enzyme inhibitor derived from an enzymatic hydrolysate of casein. II. Isolation and bradykinin-potentiating activity on the uterus and the ileum of rats. *Agric. Biol. Chem.* **1985**, *49*, 1405–1409.
16. Maruyama, S.; Suzuki, H. A peptide inhibitor of angiotensin I-converting enzyme in the tryptic hydrolysate of casein. *Agric. Biol. Chem.* **1982**, *46*, 1393–1394.
17. Maeno, M.; Yamamoto, N.; Takano, T. Identification of an antihypertensive peptide from casein hydrolysate produced by a proteinase from Lactobacillus helveticus CP790. *J. Dairy Sci.* **1996**, *79*, 1316–1321. [CrossRef]

18. Migliore-Samour, D.; Floch, F.; Jollès, P. Biologically active casein peptides implicated in immunomodulation. *J. Dairy Res.* **1989**, *56*, 357–362. [CrossRef] [PubMed]

19. Kitazawa, H.; Yonezawa, K.; Tohno, M.; Shimosato, T.; Kawai, Y.; Saito, T.; Wang, J.M. Enzymatic digestion of the milk protein beta-casein releases potent chemotactic peptide(s) for monocytes and macrophages. *Int. Immunopharmacol.* **2007**, *7*, 1150–1159. [CrossRef] [PubMed]

20. Hayes, M.; Stanton, C.; Slattery, H.; O'Sullivan, O.; Hill, C.; Fitzgerald, G.F.; Ross, R.P. Casein fermentate of Lactobacillus animalis DPC6134, contains a range of novel propeptide angiotensin-converting enzyme inhibitors. *Appl. Environ. Microbiol.* **2007**, *73*, 4658–4667. [CrossRef] [PubMed]

21. Sato, R.; Shindo, M.; Gunshin, H.; Noguchi, T.; Naito, H. Characterization of phosphopeptide derived from bovine beta-casein: An inhibitor to intra-intestinal precipitation of calcium phosphate. *Biochim. Biophys. Acta* **1991**, *1077*, 413–415. [CrossRef]

22. Loukas, S.; Varoucha, D.; Zioudrou, C.; Streaty, R.A.; Klee, W.A. Opioid activities and structures of .alpha.-casein-derived exorphins. *Biochemistry* **1983**, *22*, 4567–4573. [CrossRef] [PubMed]

23. Karaki, H.; Doi, K.; Sugano, S.; Uchiwa, H.; Sugai, R.; Murakami, U.; Takemoto, S. Antihypertensive effect of tryptic hydrolysate of milk casein in spontaneously hypertensive rats. *Comp. Biochem. Physiol. C* **1990**, *96*, 367–371. [PubMed]

24. Minervini, F.; Algaron, F.; Rizzello, C.G.; Fox, P.F.; Monnet, V.; Gobbetti, M. Angiotensin I-converting-enzyme-inhibitory and antibacterial peptides from Lactobacillus helveticus PR4 proteinase-hydrolyzed caseins of milk from six species. *Appl. Environ. Microbiol.* **2003**, *69*, 5297–5305. [CrossRef] [PubMed]

25. Srinivas, S.; Prakash, V. Bioactive peptides from bovine milk alpha-casein: Isolation, characterization and multifunctional properties. *Int. J. Pept. Res. Ther.* **2010**, *16*, 7–15. [CrossRef]

26. Yoshikawa, M.; Tani, F.; Ashikaga, T.; Yoshimura, T.; Chiba, H. Purification and Characterization of an Opioid Antagonist from a Peptic Digest of Bovine κ-Casein. *Agric. Biol. Chem.* **1986**, *50*, 2951–2954. [CrossRef]

27. Chiba, H.; Tani, F.; Yoshikawa, M. Opioid antagonist peptides derived from kappa-casein. *J. Dairy Res* **1989**, *56*, 363–366. [CrossRef] [PubMed]

28. Jollès, P.; Lévy-Toledano, S.; Fiat, A.M.; Soria, C.; Gillessen, D.; Thomaidis, A.; Dunn, F.W.; Caen, J.P. Analogy between fibrinogen and casein. Effect of an undecapeptide isolated from kappa-casein on platelet function. *Eur. J. Biochem.* **1986**, *158*, 379–382. [CrossRef] [PubMed]

29. Patten, G.S.; Head, R.J.; Abeywardena, M.Y. Effects of casoxin 4 on morphine inhibition of small animal intestinal contractility and gut transit in the mouse. *Clin. Exp. Gastroenterol.* **2011**, *4*, 23–31. [CrossRef] [PubMed]

30. Nurminen, M.L.; Sipola, M.; Kaarto, H.; Pihlanto-Leppälä, A.; Piilola, K.; Korpela, R.; Tossavainen, O.; Korhonen, H.; Vapaatalo, H. Alpha-lactorphin lowers blood pressure measured by radiotelemetry in normotensive and spontaneously hypertensive rats. *Life Sci.* **2000**, *66*, 1535–1543. [CrossRef]

31. Mullally, M.M.; Meisel, H.; FitzGerald, R.J. Synthetic peptides corresponding to alpha-lactalbumin and beta-lactoglobulin sequences with angiotensin-I-converting enzyme inhibitory activity. *Biol. Chem. Hoppe Seyler* **1996**, *377*, 259–260. [PubMed]

32. FitzGerald, R.J.; Meisel, H. Lactokinins: Whey protein-derived ACE inhibitory peptides. *Nahrung* **1999**, *43*, 165–167. [CrossRef]

33. Mullally, M.M.; Meisel, H.; FitzGerald, R.J. Identification of a novel angiotensin-I-converting enzyme inhibitory peptide corresponding to a tryptic fragment of bovine beta-lactoglobulin. *FEBS Lett.* **1997**, *402*, 99–101. [CrossRef]

34. Yamauchi, R.; Ohinata, K.; Yoshikawa, M. Beta-lactotensin and neurotensin rapidly reduce serum cholesterol via NT2 receptor. *Peptides* **2003**, *24*, 1955–1961. [CrossRef] [PubMed]

35. Tani, F.; Shiota, A.; Chiba, H.; Yoshikawa, M. Serophin, an opioid peptide derived from serum albumin. In β-*Casomorphins and Related Peptides: Recent Developments*; Brantl, V., Teschemacher, H., Eds.; Wiley: Weinheim, Germany, 1994; pp. 49–53.

36. Meisel, H.; Bockelmann, W. Bioactive peptides encrypted in milk proteins: Proteolytic activation and thropho-functional properties. *Antonie Van Leeuwenhoek* **1999**, *76*, 207–215. [CrossRef] [PubMed]

37. Meisel, H. Multifunctional peptides encrypted in milk proteins. *Biofactors* **2004**, *21*, 55–61. [CrossRef] [PubMed]

38. Fiat, A.M.; Jolles, P. Caseins of various origins and biologically active casein peptides and oligosaccharides: Structural and physiological aspects. *Mol. Cell. Biochem.* **1989**, *87*, 5–30. [CrossRef] [PubMed]
39. Meisel, H. Overview on milk protein-derived peptides. *Int. Dairy J.* **1998**, *8*, 363–373. [CrossRef]
40. Meisel, H.; FitzGerald, R.J. Opioid peptides encrypted in milk proteins. *Br. J. Nutr.* **2000**, *84*, S27–S31. [CrossRef] [PubMed]
41. Vermeirssen, V.; van Camp, J.; Verstraete, W. Bioavailability of angiotensin I-converting enzyme inhibitory peptides. *Br. J. Nutr.* **2004**, *92*, 357–366. [CrossRef] [PubMed]
42. Ganapathy, V.; Brandsch, M.; Leibach, F.H. Intestinal transport of amino acids and peptides. In *Physiology of the Gastrointestinal Tract*; Johnson, L.R., Ed.; Raven Press Ltd.: New York, NY, USA, 1994; pp. 1773–1794.
43. Pauletti, G.M.; Gangwar, S.; Knipp, G.T.; Nerurkar, M.M.; Okumu, F.W.; Tamura, K.; Siahaan, T.J.; Borchardt, R.T. Structural requirements for intestinal absorption of peptide drugs. *J. Control. Release* **1996**, *41*, 3–17. [CrossRef]
44. Pauletti, G.M.; Okumu, F.W.; Borchardt, R.T. Effect of size and charge on the passive diffusion of peptides across Caco-2 cell monolayers via the paracellular pathway. *Pharm. Res.* **1997**, *14*, 164–168. [CrossRef] [PubMed]
45. Cardillo, G.; Gentilucci, L.; Tolomelli, A.; Calienni, M.; Qasem, A.R.; Spampinato, S. Stability against enzymatic hydrolysis of endomorphin-1 analogues containing beta-proline. *Org. Biomol. Chem.* **2003**, *1*, 1498–1502. [CrossRef] [PubMed]
46. Mizuno, S.; Nishimura, S.; Matsuura, K.; Gotou, T.; Yamamoto, N. Release of short and proline-rich antihypertensive peptides from casein hydrolysate with an Aspergillus oryzae protease. *J. Dairy Sci.* **2004**, *87*, 3183–3188. [CrossRef]
47. Savoie, L.; Agudelo, R.A.; Gauthier, S.F.; Marin, J.; Pouliot, Y. In vitro determination of the release kinetics of peptides and free amino acids during the digestion of food proteins. *J. AOAC Int.* **2005**, *88*, 935–948. [PubMed]
48. Fujita, H.; Yoshikawa, M. LKPNM: A prodrug-type ACE-inhibitory peptide derived from fish protein. *Immunopharmacology* **1999**, *44*, 123–127. [CrossRef]
49. Svedberg, J.; de Hass, J.; Leimenstoll, G.; Paul, F.; Teschemacher, H. Demonstration ofmbeta-casomorphin immunoreactive materials in in vitro digests of bovine milk and in small intestine contents after bovine milk ingestion in adult humans. *Peptides* **1985**, *6*, 825–830. [CrossRef]
50. Matar, C.; Amiot, J.; Savoie, L.; Goulet, J. The effect of milk fermentation by Lactobacillus helveticus on the release of peptides during in vitro digestion. *J. Dairy Sci.* **1996**, *79*, 971–979. [CrossRef]
51. Miquel, E.; Gomez, J.A.; Alegria, A.; Barbera, R.; Farre, R.; Recio, I. Identification of casein phosphopeptides released after simulated digestion of milk-based infant formulas. *J. Agric. Food Chem.* **2005**, *53*, 3426–3433. [CrossRef] [PubMed]
52. Nongonierma, A.B.; FitzGerald, R.J. The scientific evidence for the role of milk protein-derived bioactive peptides in humans: A review. *J. Funct. Foods* **2015**, *17*, 640–656. [CrossRef]
53. Hartmann, R.; Meisel, H. Food-derivedpeptides with biological activity: From research to food applications. *Curr. Opin. Biotechnol.* **2007**, *18*, 1–7. [CrossRef] [PubMed]
54. Capriotti, A.L.; Cavaliere, C.; Piovesana, S.; Samperi, R.; Laganà, A. Recent Trends in the analysis of bioactive peptides in milk and dairy products. *Anal. Bioanal. Chem.* **2016**, *408*, 2677–2685. [CrossRef] [PubMed]
55. Muro Urista, C.; Álvarez Fernández, R.; Riera Rodriguez, F.; Arana Cuenca, A.; Téllez Jurado, A. Review: Production and functionality of active peptides from milk. *Food Sci. Technol. Int.* **2011**, *17*, 293–317. [CrossRef] [PubMed]
56. Dallas, D.C.; Lee, H.; Parc, A.L.; de Moura Bell, J.M.L.N.; Barile, D. Coupling Mass Spectrometry-Based "Omic" Sciences with Bioguided Processing to Unravel Milk's Hidden Bioactivities. *Adv. Dairy Res.* **2013**, *1*, 104. [CrossRef]
57. Sánchez-Rivera, L.; Martínez-Maqueda, D.; Cruz-Huerta, E.; Miralles, B.; Recio, I. Peptidomics for discovery, bioavailability and monitoring of dairy bioactive peptides. *Food Res. Int.* **2014**, *63*, 170–181. [CrossRef]
58. Kareb, O.; Gomaa, A.; Champagne, C.P.; Jean, J.; Aïder, M. Electro-activation of sweet defatted whey: Impact on the induced Maillard reaction products and bioactive peptides. *Food Chem.* **2017**, *221*, 590–598. [CrossRef] [PubMed]
59. Pihlanto-Leppälä, A. Bioactive peptides derived from bovine whey proteins: Opioid and ace-inhibitory peptides. *Trends Food Sci. Technol.* **2000**, *11*, 347–356. [CrossRef]

60. Bitri, L. Optimization Study for the Production of an Opioid-like Preparation from Bovine Casein by Mild Acidic Hydrolysis. *Int. Dairy J.* **2004**, *14*, 535–539. [CrossRef]
61. Hafeez, Z.; Cakir-Kiefer, C.; Roux, E.; Perrin, C.; Miclo, L.; Dary-Mourot, A. Strategies of producing bioactive peptides from milk proteins to functionalize fermented milk products. *Food Res. Int.* **2014**, *63*, 71–80. [CrossRef]
62. Linares, D.M.; Gómez, C.; Renes, E.; Fresno, J.M.; Tornadijo, M.E.; Ross, R.P.; Stanton, C. Lactic Acid Bacteria and Bifidobacteria with Potential to Design Natural Biofunctional Health-Promoting Dairy Foods. *Front. Microbiol.* **2017**, *8*. [CrossRef] [PubMed]
63. Mohanty, D.P.; Mohapatra, S.; Misra, S.; Sahu, P.S. Milk derived bioactive peptides and their impact on human health—A review. *Saudi J. Biol. Sci.* **2016**, *23*, 577–583. [CrossRef] [PubMed]
64. Park, Y.W.; Nam, M.S. Bioactive Peptides in Milk and Dairy Products: A Review. *Korean J. Food Sci. Anim. Resour.* **2015**, *35*, 831–840. [CrossRef] [PubMed]
65. Krissansen, G.W. Emerging health properties of whey proteins and their clinical implications. *J. Am. Coll. Nutr.* **2007**, *26*, 713S–723S. [CrossRef] [PubMed]
66. Korhonen, H.; Pihlanto, A. Food-derived bioactive peptides—Opportunities for designing future foods. *Curr. Pharm. Des.* **2003**, *9*, 1297–1308. [CrossRef] [PubMed]
67. Korhonen, H.; Marnila, P. Bovine milk antibodies for protection against microbial human diseases. In *Nutraceutical Proteins and Peptides in Health and Disease*; Mine, Y., Shahidi, S., Eds.; Taylor & Francis Group: Boca Raton, FL, USA, 2005; pp. 137–159.
68. FitzGerald, R.J.; Murray, B.A.; Walsh, D.J. Hypotensive peptides from milk proteins. *J. Nutr.* **2004**, *134*, 980–988.
69. Dziuba, B.; Dziuba, M. Milk proteins-derived bioactive peptides in dairy products: Molecular, biological and methodological aspects. *Acta Sci. Pol. Technol. Aliment.* **2014**, *13*, 5–25. [CrossRef] [PubMed]
70. Korhonen, H.; Pihlanto, A. Bioactive peptides: Production and functionality. *Int. Dairy J.* **2006**, *16*, 945–960. [CrossRef]
71. Brandelli, A.; Daroit, D.J.; Folmer Corrêa, A.P. Whey as a source of peptides with remarkable biological activities. *Food Res. Int.* **2015**, *73*, 149–161. [CrossRef]
72. Yadav, J.S.; Yan, S.; Pilli, S.; Kumar, L.; Tyagi, R.D.; Surampalli, R.Y. Cheese whey: A potential resource to transform into bioprotein, functional/nutritional proteins and bioactive peptides. *Biotechnol. Adv.* **2015**, *33*, 756–774. [CrossRef] [PubMed]
73. Sommella, E.; Pepe, G.; Ventre, G.; Pagano, F.; Conte, G.M.; Ostacolo, C.; Manfra, M.; Tenore, G.; Russo, M.; Novellino, E.; et al. Detailed peptide profiling of "Scotta": From a dairy waste to a source of potential health-promoting compounds. *Dairy Sci. Technol.* **2016**, *96*, 763–771. [CrossRef]
74. Patel, S. Emerging trends in nutraceutical applications of whey protein and its derivatives. *J. Food Sci. Technol.* **2015**, *52*, 6847–6858. [CrossRef] [PubMed]
75. Athira, S.; Mann, B.; Saini, P.; Sharma, R.; Kumar, R.; Singh, A.K. Production and characterisation of whey protein hydrolysate having antioxidant activity from cheese whey. *J. Sci. Food Agric.* **2015**, *95*, 2908–2915. [CrossRef] [PubMed]
76. Abd El-Salam, M.H.; El-Shibiny, S. Preparation, properties, and uses of enzymatic milk protein hydrolysates. *Crit. Rev. Food Sci. Nutr.* **2017**, *57*, 1119–1132. [CrossRef] [PubMed]

Review

Role of Proteins and of Some Bioactive Peptides on the Nutritional Quality of Donkey Milk and Their Impact on Human Health

Silvia Vincenzetti [1,*], Stefania Pucciarelli [1], Valeria Polzonetti [1] and Paolo Polidori [2]

[1] School of Bioscience and Veterinary Medicine, University of Camerino, via Gentile III da Varano, Camerino (MC) 62032, Italy; stefania.pucciarelli@unicam.it (S.P.); valeria.polzonetti@unicam.it (V.P.)

[2] School of Pharmacy, University of Camerino, via Circonvallazione 93, Matelica (MC) 62024, Italy; paolo.polidori@unicam.it

* Correspondence: silvia.vincenzetti@unicam.it; Tel.: +39-073-740-2722; Fax: +39-073-740-3427

Academic Editor: Alessandra Durazzo

Received: 25 May 2017; Accepted: 5 July 2017; Published: 10 July 2017

Abstract: Donkey milk could be considered a good and safer alternative, compared to other types of milk, for infants affected by cow's milk protein allergy, when breastfeeding is not possible. Interestingly, donkey milk has low allergenicity, mainly due to the low total casein amount, and the content of some whey proteins that act as bioactive peptides. The amount of lysozyme, an antibacterial agent, is 1.0 g/L, similar to human milk. Lactoferrin content is 0.08 g/L, with this protein being involved in the regulation of iron homoeostasis, anti-microbial and anti-viral functions, and protection against cancer development. Lactoperoxidase, another protein with antibacterial function, is present in donkey milk, but in very low quantities (0.11 mg/L). β-lactoglobulin content in donkey milk is 3.75 g/L—this protein is able to bind and transport several hydrophobic molecules. Donkey milk's α-lactalbumin concentration is 1.8 g/L, very close to that of human milk. α-lactalbumin shows antiviral, antitumor, and anti-stress properties. Therefore, donkey milk can be considered as a set of nutraceuticals properties and a beverage suitable, not only for the growing infants, but for all ages, especially for convalescents and for the elderly.

Keywords: donkey milk; bioactive peptide; whey proteins; nutritional properties

1. Introduction

1.1. An Overview on Milk Composition

Milk is a natural beverage which meets the nutritional needs of infants, since it is one of the most complete and highly nutritious foods. Nutritionally, milk is defined as "the most nearly perfect food" due to its chemical composition. The principal constituents are water, lipids, carbohydrates, and proteins (caseins and whey proteins). Furthermore, other minor constituents are present in milk, such as vitamins, minerals, hormones, enzymes, and miscellaneous compounds. As well, there are numerous immune components like growth factors, cytokines, nucleotides, antimicrobial compounds, and specific immune cells. Thanks to these features, milk provides important nutritive elements, immunological protection, and biologically active substances. The principal milk constituents vary widely among species: lipids, less than 1–55%; proteins, 1–20%; lactose, 0–10%. Concentrations of the minor constituents are also variable. Within any species, milk composition changes among individual animals, between breeds, according to the stage of lactation, feed and health of the animal, the environment, and the climatic situation, among many other factors [1].

The composition of milk reflects mainly the nutritional and physiological requirements of the newborn, and even the profile of constituents therein changes markedly during lactation. The main changes are evident during the first days post-partum, especially regarding the immunoglobulin fraction. The concentration of milk components remains relatively constant during mid-lactation and changes noticeably in late lactation, due to the involution of the mammary gland tissue and the greater influx of blood constituents [2].

Among the milk constituents, fats vary widely between different species both in their concentration and chemical composition, from less than 1% in donkey milk, to more than 50% in aquatic mammals [3]. Its composition depends on the energy requirements and on the nutritional, genetic, and lactation characteristics of the species. Milk lipids are also important as a source of essential fatty acids, such as linoleic and linolenic acid, and fat-soluble vitamins A, D, E, and K.

Considering this milk fraction, although 97–98% of lipids are triacylglycerols, small amounts of di- and monoacylglycerols, free cholesterol and cholesterol esters, free fatty acids, and phospholipids are also present. Phospholipids, which represent less than 1% of total lipid, are present mainly in the milk fat globule membrane and other membranous material in the milk [4].

Regarding carbohydrate content in milk, lactose concentration varies from 0.7% to about 7.0% among the different mammal species [4].

Lactose is synthesized in the epithelial mammary cells from two molecules of glucose: one molecule of glucose is converted into UDP-galactose, which is condensed with a second molecule of glucose by the complex of lactose synthetase. The latter is a dimer composed by the UDP-galactosyl transferase and the whey protein α-lactalbumin (α-La), which is a regulatory subunit able to make UDP-galactosyl transferase specific for glucose. Therefore, there is a positive correlation between the concentrations of lactose and α-La in milk, (the milk of the Californian sea lion, which contains no lactose, also lacks α-La). α-La probably has a regulatory role in the lactose synthesis, as well as in the control of the osmotic pressure of milk [5].

Milk protein content ranges from 1% to 24% by milk weight depending on the mammalian species. There are two major categories of milk proteins that are defined by their chemical composition and physical properties: caseins and whey proteins. The caseins, which accounts for about the 80% of milk proteins in ruminants, are responsible for the transport of calcium and phosphate, and for the formation of a clot in the stomach for efficient digestion.

They are heat stable phosphoproteins synthesized in the mammary glands and found in the milk of all mammals. Caseins are divided in α-caseins, which in turn comprise αs_1-, αs_2-, β-, and κ-caseins. They differ in primary structure and type and degree of post-translational modifications. These proteins are able to bind and transport relatively large amount of calcium, thanks to the presence of phosphates: this makes caseins very important for the development of teeth and bones in newborns [6].

The casein proteins are nutritionally important also because of their high content in lysine, an essential amino acid in humans: αs_1- and αs_2-casein have 14 and 24 lysine residues, respectively.

The whey proteins are those proteins which do not precipitate from solution when the pH of milk is adjusted to 4.6. These proteins are referred to also as serum proteins or non-casein nitrogen, and are a heterogeneous group of heat labile globular proteins constituted mainly by α-lactalbumin (α-La), β-lactoglobulin (β-LG), serum albumin and immunoglobulins, and to a lesser extent by lactoferrin and lysozyme. In the whey protein fraction, there are also enzymes, hormones, nutrient transporters, growth factors, disease resistance factors, and others. However, contrary to caseins, they do not contain phosphorus.

The ratio of casein/whey proteins is different among species—in human milk, the ratio is about 40/60, in mare's milk it is around 50/50, while in cow, goat, sheep, and buffalo's milk it is about 80/20. These differences reflect the nutritional and physiological requirements of the newborn of these species [7].

The inorganic elements are defined as macroelements or microelements, depending on whether they are present in the animal organism in smaller or greater quantities. The most important mineral salts in milk are calcium, sodium, potassium, magnesium, phosphorus (inorganic), chlorides, and citrates. Mineral elements are indispensable for the life of animals, as they participate in numerous physiological functions.

In milk, the inorganic elements are in solution or complexed into casein micelles.

In Tables 1 and 2 are reported respectively, the quantities of the main macroelements and microelements in the milk of various species.

Table 1. Concentration of inorganic elements (mM) in the milk of different species (adapted from [8]).

	Human	Bovine	Caprine	Mare	Swine	Ovine
Calcium	7.8	29.4	23.1	16.5	104.1	56.8
Sodium	5.0	24.2	20.5	5.7	14.4	20.5
Potassium	16.5	34.7	46.6	11.9	31.4	31.7
Magnesium	1.1	5.1	5.0	1.6	9.6	9.0
Phosphorus	2.5	20.9	15.6	6.7	51.2	39.7
Chloride	6.2	30.2	34.2	6.6	28.7	17.0

Table 2. Concentration of some trace elements (μM) in mare, bovine, and human milk (adapted from [9]).

	Bovine	Mare	Human
Zinc	3960	1835	2150
Iron	194	224	260
Copper	52	155	314
Manganese	21	14	7
Barium	188	76	149
Aluminium	98	123	125

Vitamins are important bioregulators and are classified as fat-soluble and water-soluble. They are essential compounds for nutrition as they are involved in several functions: some are hormone precursors (i.e., vitamin D), others act as antioxidants by sequestering free radicals (i.e., vitamins C and E), and others constitute the prosthetic group of enzymes (i.e., vitamin B complex). Vitamin nutritional requirement varies from a few mg, to a few μg per day, depending on the considered vitamin. In milk, vitamin content depends on the maternal vitamin status and, especially for the water-soluble vitamins, from the maternal diet. In Table 3, the content of some fat-soluble and water-soluble vitamins in the milk of various species is summarized.

Table 3. Vitamin content (mg/L) in the milk of different species (adapted from [10]).

Fat-Soluble Vitamins	Bovine	Caprine	Mare	Human	Ovine
A and β-carotene	0.32–0.50	0.50	0.12	2.0	0.50
D_3, Cholecalciferol	0.003	-	0.003	0.001	-
E, α-tocopherol	0.98–1.28	-	1.13	6.60	-
K	0.011	-	0.020	0.002	-
Water-soluble vitamins					
B_1, thiamine	0.37	0.49	0.30	0.15	0.48
B_2, riboflavin	1.80	1.50	0.30	0.38	2.30
B_3, niacin	0.90	3.20	1.40	1.70	4.50
B_5, pantothenic acid	3.50	3.10	3.0	2.70	3.50
B_6, pyridoxine	0.64	0.27	0.3	0.14	0.27
B_7, biotin	0.035	0.039	-	0.006	0.09
B_9, folic acid	0.18	n.d	n.d	0.16	n.d
B_{12}, cobalamin	0.004	0.70	0.003	0.5	0.007
C, ascorbic acid	21.0	9.0	17.2	43.0	4.25

1.2. Cow's Milk Protein Allergy

It is well-known that the prevalence of adverse reactions to food is higher in children (6–8% in the early years of life) than in adults (2.4%) [11]. Prospective studies, performed in various countries, showed that about 2.5% of children in the first year of life have allergic reactions to cow's milk, which is therefore considered the main cause of allergies in children [12].

Cow's Milk Protein Allergy (CMPA) is clinically an abnormal immunological reaction to cow milk proteins. It may be due to the interaction between one or more milk proteins and one or more immune mechanisms, and resulting in IgE-mediated reactions. If the reactions do not involve the immune system, they are defined as cow milk protein intolerance. IgE-mediated forms are often associated with a higher risk of multiple food allergies and atopic conditions, such as asthma in later periods of life. The causes of allergy to cow's milk are manifold: first, the protein content of cow's milk is higher than that of human milk (3.6% versus 0.9%–1%); second, cow's milk proteins are represented by casein for the 80% and by the whey proteins for the 18% (in human milk, 40% caseins and 60% whey proteins); third, β-lactoglobulin is absent in human milk, and for this reason, this protein is considered as a potential allergen (allergen Bos d 5).

Cow's milk caseins (αs_1-, αs_2-, β- and κ-casein) constitute the allergen Bos d 8, however, there is a greater sensitiveness to the α-caseins (100%) and the κ-casein (91.7%). Among whey proteins, β-LG is the most abundant whey protein in cow's milk and it affects allergic patients in a percentage from 13% to 76%. Also, bovine immunoglobulins (allergen Bos d 7), and in particular γ-globulin, are responsible for CMPA clinical symptoms [13]. However, the allergic reaction to cow's milk proteins occurs when these proteins are poorly digested by the infant and hence their structure remains substantially unchanged, since the polypeptide chain is not fractionated into amino acids because of insufficient protease enzymatic attack, or a lack of a specific intestine enzyme. Therefore, the undigested protein behaves as an antigen. This is not enough to determine allergic reactions, since the amount of absorbed antigen is limited by anatomical barriers (mucus, epithelium) and/or eliminated by immunological barriers, by a combination on the epithelial surface of antigen with secretory immune globulin (IgA). However, if the anatomic barriers are weakened, because of inflammatory or IgA deficiency, the penetration of large amounts of antigens occurs and the awareness and allergy in atopic subjects could increase. CMPA occurs with gastrointestinal disorders for 50–60% of cases, skin diseases for 50–70%, respiratory affections for 20–30% respirators, and anaphylaxis for 5–9% of cases [14].

Although milk allergy is considered transient, there are still some children who do not outgrow their allergy by the age of 10, but also by their adulthood [15]. These latter patients have more severe reactions and a different immunological pattern than those with transient allergy [16].

Once CMPA has been established, the therapy, as in all food allergies, should consist of the absolute exclusion of milk or allergenic proteins of the cow's milk (formulated milk and derivatives), whereas in the case of breastfed children, on the complete elimination of milk protein from the mother's diet [17].

When breast milk is insufficient or unavailable, it is indispensable to choose an alternative milk formula. In the case of infants affected by CMPA, the elimination of allergen administration usually leads to the disappearance of the symptoms. However, some patients may also show the allergic reaction if a substitute for cow's milk is used, thus indicating multiple food allergies [18].

Multiple food allergies are difficult to treat—elimination from the diet of some nutrients, if not properly balanced by the assumption of alternative nutrients, can lead to malnutrition and therefore can stop growing [18].

2. Donkey Milk and Its Hypoallergenic Properties

Donkey milk was greatly appreciated for its benefits and versatility as a food for newborns since ancient times, but only recently, scientific research has proven its importance in human nutrition. This is especially for some categories of the population as the elderly and the infants affected by CMPA.

2.1. Historia Docet

In ancient Egypt, as evidenced by the images of bas-reliefs and painting from that time, donkeys were very important for their use in agriculture and for daily works (Figure 1).

Figure 1. A bas-relief and a panting from ancient Egypt showing the use of donkeys for daily works.

The ancient Greeks considered donkey milk an excellent drug. In the 5th century BC, Herodotus estimated it as a nutritious beverage, whereas Hippocrates recommended it as a medicine capable of resolving several afflictions. Ancient Romans consumed donkey milk as a delicious drink and Pliny the Elder described its virtues regarding skin wellness: "It is believed that the donkey's milk eliminates skin wrinkles and makes the face softer and whiter. Some women are known to care for their face seven times a day, paying close attention to this number." For this purpose, it is well-known that Cleopatra and Poppaea used to immerse themselves in donkey milk to preserve their youth. In particular, Poppaea, the wife of Nero, used donkey milk every day for the bath and for this reason, she carried back 500 donkeys every time she left for a trip. It is said that Messalina also loved these "baths of beauty" for the anti-wrinkle action that she guaranteed on the skin [19]. In Asia, Africa, and Europe, it is possible to find testimonies and documents describing the properties of donkey milk, decanted for its therapeutic, cosmetic, and food virtues. In Russia and Mongolia, the consumption of milk and donkey milk, thanks to the abundant presence of vitamins A, B, and particularly C, compensated for the low consumption of fruit, vegetables, and legumes. Nomads also used to have several donkeys and horses for their milk. Leo Tolstoy affirmed: "Mare's milk give vigor to my body and wings to my spirit". During the Renaissance, there was the first scientific consideration of this product, when Francis I of France, based on the suggestion of his doctors, successfully used donkey milk against stress and physical fatigue. In the nineteenth century, again in France, Dr. Parrot of the "Hôpital des Enfants Assistés", breastfed orphaned babies directly to the nipple of the donkey (Figure 2); moreover, in several European cities, it was possible to meet some merchants who sold donkey milk.

Figure 2. Collecting Ass' milk. Hôpital St. Vincent de Paul. Paris XIX century.

In those years, therefore, there was consciousness raising about the validity of donkey milk as a substitute for human milk, using it in nursing infants whenever the case requires it. This practice has been in use until the 1950s. In fact, during these years, it was normal to use donkey milk in cases where there was a shortage of breast milk, or when intolerance towards cow's milk was observed.

Thus, nowadays, it important to take into account the many testimonies in the past on the beneficial effects of donkey milk, but it is also important to give the right scientific approach that attests and supports its innumerable properties that are here summarized: strengthens the immune system, acts as a prebiotic on the gut flora, calms laryngeal irritation and cough, fights anaemia, fights skin problems such as psoriasis, acne, and eczema, acts on skin disorders caused by stress, and is highly digestible by newborns [20]. For these reasons, interest in donkey milk has recently begun, pressing medical research to expand its literature on this topic. On the other hand, the scientific community inherits a knowledge from the historical tradition of its great value—in particular, the importance of donkey milk as the most suitable for infant feeding as it is the closest to human milk, its high nutritional qualities, and high digestibility.

2.2. Donkey Milk and Its Affinity with Human Milk

It is well-known that among mammals, every species produces a kind of milk that is specific to their newborns, ensuring them the optimal nutrition. This is also true for humans. The best milk for an infant is breast milk, which responds perfectly to the needs of healthy growth and harmonious development. However, when it is not possible to breastfeed, and especially in the case of CMPA, it is necessary to find a safe and valid alternative milk that meets the needs of the infant. The various kinds of alternative milk are certainly valid and ensure good child growth, however, sometimes they are characterized by an unpleasant taste and a high cost. Further, in the case of multiple food allergies, may themselves cause unwanted allergic reactions. Donkey milk is considered a milk similar to human milk, especially regarding its protein composition. Therefore, as well as breast milk, it is able to respond to all the food needs of the newborn. Donkey milk is also an important food in the geriatric field and in cases of calcium deficiency [21].

Some studies have shown that donkey milk, which has the organoleptic characteristics closer to breast milk, may be the treatment of choice in children with food allergies in early life, who often do not respond to other therapies. Unlike other breast milk substitutes, characterized by nutritional deficiencies and induction of allergic reactions, this natural food proves to be able to nourish infants at low risk of allergenicity and also allows for the development of a normal and complete immune system [22]. With this in mind, we can consider donkey milk not simply as a food, but as a nutraceutical. We consider it to be used not only in early childhood nutrition, but also as a supplement in the diet of adults and the elderly.

In Table 4, the average percentage composition (g/100 g) of donkey milk (evidenced in bold) is shown, compared to that of human, bovine, mare, ovine, and caprine milk [23,24]. The fat content in donkey milk is very close to that of mare milk, but low when compared to that of other species, including human milk. As a consequence, the average energy value found in donkey milk is 1939.4 kJ/kg, similar to that of mare's milk, but lower with respect to the milk of other species [25]. This low energy value could be a problem for the infant's energy intake if the diet of the latter is based only on donkey milk. However, this problem could be overcome through a supplement of medium-chain triglycerides or sunflower oil in the milk [26,27]. Regarding the lipid composition, donkey milk contains a lower amount of saturated fatty acid (SFA) and higher level of essential fatty acid (EFA) than that of bovine milk, and a high polyunsaturated fatty acids (PUFA) content (especially α-linolenic acid and linoleic acid). Furthermore, this milk is characterized by a low ratio of n-6 to n-3 fatty acids [28,29].

The total donkey milk dry matter is 9.61%, ashes corresponding approximately to 0.43%. The proportion of non-protein nitrogen is very close to that found in human milk. The biological significance of the non-protein fraction is not well clarified. However, this fraction includes urea, uric acid, creatinine, nucleic acids, amino acids, and nucleotides, which seem to be very important for neonatal development [30].

Table 4. Percentage composition (g/100 g) of donkey milk (evidenced in bold) and comparison with other species.

Milk	Water	Dry Matter	Fat	Proteins	Lactose	Ashes	Energy Value (kJ/kg)
Human	87.57	12.43	3.38	1.64	6.69	0.22	2855.6
Donkey	**90.39**	**9.61**	**1.21**	**1.74**	**6.23**	**0.43**	**1939.4**
Mare	90.48	9.52	0.85	2.06	6.26	0.35	1877.8
Bovine	87.62	12.38	3.46	3.43	4.71	0.78	2983.0
Caprine	86.77	13.23	4.62	3.41	4.47	0.73	3399.5
Ovine	80.48	19.52	7.54	6.17	4.89	0.92	5289.4

Adapted from [23,24].

Donkey milk contains a high level of lactose, which is involved in bone mineralization. Its lactose concentration, in particular, is close to that of human milk [22]. It is also responsible for the sweet taste of donkey milk, which is much more pleasing to the palate of the newborn with respect to the other breast milk substitutes. Lactose stimulates the intestinal absorption of calcium in the infant because in these subjects, the enzyme β-galactosidase (which hydrolyses lactose into galactose and glucose) is highly expressed. Some studies in fact demonstrated that β-galactosidase-deficient subjects absorbed less calcium than β-galactosidase-normal subjects because are the hydrolytic products of β-galactosidase (glucose and galactose) that promote the calcium absorption [31,32]. Therefore, even after weaning, if a balanced diet is followed, donkey milk favors bones mineralization.

Finally, the presence of lactose makes donkey milk a suitable substrate for the preparation of fermented beverages [33].

Considering donkey milk protein content, the average of total caseins and whey proteins contents were found similar to those of human milk, but lower when compared to bovine milk (Table 5).

The content of casein was found to be slightly higher in donkey milk than in human milk, but significantly lower than that of sheep and bovine milk (Table 5). Moreover, the ratio casein/whey proteins were higher in donkey milk than human milk, but were closer to that of breast milk substitution products. In contrast, in ruminant milk, this ratio is four times higher than that of donkey milk and seven times greater than that of humans [34].

From the evaluation of the sequence homologies of $αs_1$-casein and β-casein from donkey, human, and cow, the highest homology was found between the donkey and human species both for α1-casein (42% donkey/human versus 31% cow/human) than for β-CN (57% donkey/human vs. 54% cow/human) [35]. This data, strictly dependent on the amino acid sequence of the comparative proteins, together with the low quantitative level of these two fractions in the donkey milk, justifies and demonstrates the best tolerability of the milk of this species.

In order to find out more information about the nutritional properties and the hypoallergenic value of donkey milk, several studies have been conducted in recent years in order to examine the protein component of this nutritionally important beverage.

3. Bioactive Proteins and Peptides in Donkey Milk

In Table 5 are listed in detail the different protein fractions (caseins and whey proteins) identified in donkey milk, together with the respective amount (g/L). Donkey's milk presents an amount of α-La rather close to that determined in human milk a high amount of β-LG, which is completely absent in human milk. This whey protein has been indicated, as mentioned, as the major allergen of cow's milk, together with caseins.

This data could affect the hypoallergenic potential of donkey milk, although the cases where this milk has been used in the treatment of CMPA have proved its efficacy with very high percentages of resolution of the allergic problem. The hypothesis formulated by several pediatricians is that caseins are the main responsible of CMPA, while β-LG plays a much smaller role in the occurrence of this pathology.

Table 5. Main proteins content in donkey milk (evidenced in bold) and comparison with bovine, caprine, and human milk.

	[a] **Bovine (g/L)**	[b] **Donkey (g/L)**	[c] **Caprine (g/L)**	[a] **Human (g/L)**
Total protein content	32.0	**13–28**	28–32	9–15
Total caseins	27.2	**6.6**	25.0	5.6
Total whey proteins	4.5	**7.5**	6.0	8.0
αs_1-casein	10.0	**n.d.**	10.0	0.8
αs_2-casein	3.7	**n.d.**	3.0	-
β-casein	10.0	**n.d.**	11.0	4.0
κ-casein	3.5	**trace**	4.0	1.0
α-lactalbumin	1.2	**1.80**	6.0	1.9–2.6
β-lactoglobulin	3.3	**3.7**	1.2	-
Lysozyme	trace	**1.0**	trace	0.04–0.2
Lactoferrin	0.1	**0.08**	0.02–0.2	1.7–2.0
Immunoglobulins	1.0	**n.d.**	1.0	1.1
Albumin	0.4	**n.d.**	0.5	0.4

[a] [36]; [b] [37,38]; [c] [39].

Of note is the lysozyme content in the donkey milk, in a quantity significantly greater than that one determined in human milk; this peptide has a bactericidal action, since it breaks the bacterial cell wall. It is also believed that the high presence of this compound gives the donkey milk the peculiarity of preserving its organoleptic and especially microbiological characteristics for a long time unchanged.

Donkey milk whey proteins content was also checked during the lactation period [37] (Table 6). Lysozyme amount decreased linearly during lactation—this protein is a natural antimicrobial agent; therefore, its amount tends to be higher after parturition with respect to the end of lactation. This is in order to ensure to the newborn an adequate protection against bacterial agent, since in the first days of life, the immune system of the child is not mature. α-La content doubles up to the nineteenth day of lactation then stabilizes—this behavior reflects the important role of this protein in the lactose synthesis. Finally, the content of β-LG remains stable during lactation.

Table 6. Content (g/L) of lysozyme, β-LG and α-La in donkey milk collected in different stages of lactation.

Lactation Period	β-LG	α-La	Lysozyme
60	n.d.	0.81	1.34
90	4.13	1.97	0.94
120	3.60	1.87	1.03
160	3.69	1.74	0.82
190	3.60	1.63	0.76

Adapted from [37].

In the following paragraphs will be discussed in more detail the protein/peptide components of donkey milk, their potential nutraceutical function, and their impact on human health.

3.1. Casein Fraction

Donkey milk casein fraction is represented mainly by αs_1- and β-caseins, which showed a great variability due to the degree of phosphorylation and the presence of genetic variants. The composition of donkey milk caseins has been the subject of several studies performed by one-dimensional electrophoresis, two-dimensional electrophoresis, structural MS analysis, and reversed phase-HPLC [38,40,41]. The presence of αs_2-casein and k-casein was also found in donkey milk even if at small amounts [40,41]. By two-dimensional electrophoresis were found more than 14 protein spots, which were in turn found corresponding mainly to αs_1-casein and β-casein, as also evidenced by reversed phase-HPLC [37,38]. Donkey milk β-caseins showed a pI ranging from 4.63 to

4.95 and a Mr varying from 33.74 to 31.15 kDa. This heterogeneity is due to the presence of two genetic variants of β-casein: a full length one characterized by the presence of 7, 6, 5 phosphate groups and pI range of 4.74–4.91 and a spliced variant (-923 amino acids) carrying 7, 6, 5 phosphate groups and a pI range of 4.72–4.61. By two-dimensional electrophoresis experiments were found five αs_1-casein in donkey milk: three of them showed similar molecular weight (about 31.15 kDa) but different pIs (range 5.15–5.36), the other two showed a lower molecular weight (27–28 kDa) and pI of 5.08 and 4.92 [38]. Also in the case of αs_1-casein the heterogeneity is due to the presence of 5, 6 and 7 phosphate groups and non-allelic spliced forms [41]. These latter authors found in donkey milk the presence of at least four αs_1-casein main components, six β-casein components, three αs_2-casein components (containing 10, 11, and 12 phosphate groups), and eleven κ-caseins (evidenced after specific immunostaining). In the case of κ-casein, this high heterogeneity is probably due to a different glycosylation pattern [41].

In Table 7 are summarized the casein component until found in donkey milk.

Table 7. Donkey milk casein components. The data reported in this table were obtained from: [38,40,41].

Casein	Mr (kDa)	pI
β-casein (full-length)	33.74	4.63
β-casein (full-length)	33.54	4.72
β-casein (full-length)	33.10	4.82
β-casein (full-length)	33.54	4.92
β-casein (spliced)	31.66	4.68
β-casein (spliced)	31.48	4.80
β-casein (spliced)	32.15	4.88
β-casein (spliced)	31.15	4.95
αs_1-casein (full-length)	31.20	5.15
αs_1-casein (full-length)	31.14	5.23
αs_1-casein (full-length)	31.14	5.36
αs_1-casein (spliced)	28.26	5.08
αs_1-casein (spliced)	27.24	4.92
αs_2-casein	26.83	n.d.
αs_2-casein	26.91	n.d.
αs_2-casein	26.99	n.d.
Eleven κ-caseins	n.d.	n.d.

3.2. Whey Protein Fraction and Its Impact on Human Health

Whey protein fraction has been well characterized in donkey milk by several authors. Some proteins possess important nutraceutical properties and may be beneficial for human health.

Hereafter, the main milk whey proteins are described separately and in detail.

3.2.1. β-Lactoglobulin

β-lactoglobulin (β-LG) is the most represented whey protein in ruminants and is also present in milk of some but not all species. It is not found in the milk of rodent, human, and lagomorph. In donkey milk β-LG content is 3.75 g/L, a value very close to that found in bovine milk (3.3 g/L) and mare milk. A proteomic study was performed by two-dimensional electrophoresis, evidenced in at least three main isoforms of donkey milk β-LG which differ in relative molecular mass (M_r) and pI value (M_r/pI: 21.9/4.46; 20.0/4.40; 20.58/4.12) [38] (Figure 3).

Other authors revealed that in donkey milk β-LG exists in two isoforms: β-lactoglobulin I (β-LGI the major form, 80%), and β-lactoglobulin II (β-LGII, present in minor amount) [42,43]. Other studies found the presence of one genetic variant of β-LGI (β-LGIB), two genetic variants of β-LGII (namely β-LGIIB and β-LGIIC), and a third minor β-LGII variant (β-LGIID) [44,45]. β-LG is a small protein with 162 amino acid residues (Mr ~18.4 kDa)—its sequence and its three-dimensional structure (it is a dimer) show that it is part of the family of lipocalins which includes a large and diversified group of over 50 extracellular proteins that originate from a large variety of tissues in animals, plants, and bacteria. A typical lipocalin consists of a peptide chain of 160–180 amino acids, folded into 8 antiparallel filaments that produce a β-sheet arranged in a conical structure called β-barrel, in which is located the hydrophobic pocket, able to bind different hydrophobic molecules. In fact, members of this family are characterized by some common molecular properties: the ability to bind several small hydrophobic molecules, the ability to bind to specific cell surface receptors, and the ability to form complexes with soluble macromolecules. For these reasons, lipocalins can act as specific transporters, for example, for serum retinol-binding protein (RBP) [46]. The high affinity of β-lactoglobulin for a wide range of compounds and its high amount in milk has suggested several functions for this protein. In addition to transporting hydrophobic molecules, this protein seems to be involved in enzyme regulation, and in the neonatal acquisition of passive immunity [46].

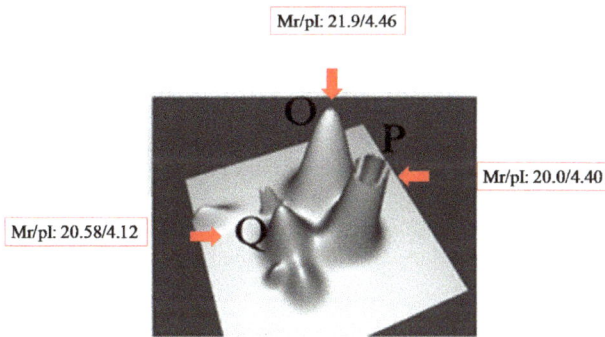

Figure 3. The three main isoforms of β-lactoglobulin, detected by two-dimensional electrophoresis. Adapted from [38].

As regard to the hydrophobic ligands, they are mainly represented by long chain fatty acids, retinol, and steroids. In milk, fatty acids, rather than retinol, represent endogenous ligands; the latter, however, being more soluble in fat, is indirectly transferred from the mother to the newborn through the β-LG. However, the effective role of this protein is not certain. This is due to two main reasons: the first is given by the presence of other lipocalins that have a similar function, the second being a simple reasoning based on the fact that if β-LG played an indispensable role, it would be expressed in all mammalian species and not just in some. One hypothesis on the role of this protein may be that the real β-LG function may be related to the physiology of the mother, rather than to that of the newborn.

An important issue has been the identification in cow and goat of a pseudo-gene (ancestral genes that have lost their ability to be expressed), which appeared very similar to the equine β-LG-II sequence [46]. An explanation of the presence of the pseudo-gene could be that the protein served for its true function elsewhere then it may be expressed in the mammary gland of some species (but not all) as source of nutrition.

Kontopidis and co-workers [46] examined the β-LG sequences of the various species together with the other lipocalins, creating a genealogical tree from which they noted that RBPs are clearly distinct from the lactoglobulins, with the exception of a single protein: the glycodelin. This is a protein expressed in a large amount during human pregnancy (first three months) in the endometrium and seems to be involved in immunosuppression and/or cellular differentiation.

Therefore, it could be possible to postulate that in many species this gene has undergone a duplication event and now it is expressed during lactation for nutritional purposes. Likewise, in some species, such as rodents, lagomorphs, and humans, the formation of a pseudogene has ceased, and consequently ceased its presence in the milk of these species [46].

Recently it was shown by Liang and co-workers [47] that β-LG is able to interact with resveratrol, a natural polyphenolic compound with antioxidant activity forming a 1:1 complex that enhances the photostability and hydrosolubility of resveratrol, improving its bioavailability. Furthermore, it has been shown that β-LG is able to form complexes also with folic acid; therefore, it may be used as an effective carrier of this important vitamin in food [48].

3.2.2. α-Lactalbumin

Donkey milk contains a considerable amount of α-La with a concentration of 1.80 g/L, very close to the amount found in human and bovine milk [37]. Thanks to two-dimensional electrophoresis it was possible to identify in donkey milk the presence of two isoforms of α-La that showed different pI values: 4.76 and 5.26 [38]. The presence of two isoforms in donkey milk was already observed by Cunsolo and co-workers [45], who found oxidized methionine (at position 90 of the amino acid sequence) form for α-La, which presumably derive by in vivo oxidative stress.

α-La is a small protein (Mr 14.2 kDa), with a binding site with Ca^{2+} ion. It has been hypothesized that α-La gene originated 300–400 million years ago from an ancestral lysozyme gene, by gene duplication. This is the reason why α-lactalbumin shares 40% sequence identity with lysozyme (also their three-dimensional structure looks very similar).

This protein consists of two domains: one large α-helical domain and another one smallest constituted by β-sheet. These two domains are linked together by calcium binding loop. Further, they are held together by a disulphide bridge between cysteine residue 73 and 91, which forms also the calcium binding loop, and another disulphide bridge between 61 and 77 cysteine residues. Definitively, the overall structure of α-La is stabilized by four disulphide bridges [49].

This protein is important for various reasons. First of all, α-La is part of the enzymatic complex that regulates lactose synthesis in the breast gland and also facilitates the absorption of lactose in the intestine. However, α-La has myriad of other functions, among all anti-inflammatory and antinociceptive functions. Some authors showed that the proteolytic digest of α-La released three peptides with bactericidal properties. In particular, two fragments were obtained from the tryptic digestion and the third one was obtained by fragmentation of α-La by chymotrypsin. These polypeptides were mainly active against Gram-positive bacteria, whereas they were less active on Gram-negative bacteria. If α-La was digested by pepsin, the polypeptide fragments obtained did not show any antibacterial activity. Since undigested α-La does not possess bactericidal activity, it was suggested as a possible antimicrobial function of α-lactalbumin only after its partial digestion by endopeptidases [50].

Other authors purified a α-La folding variant from human milk with bactericidal activity against antibiotic-resistant strains of *Streptococcus pneumoniae*. In these experiments, the native α-La was converted to the active bactericidal form by anion-exchange chromatography in the presence of a C18:1 fatty acid (oleic acid) [51]. However, this antibacterial activity was selective only for streptococci—in fact, Gram-negative and other Gram-positive bacteria resulted resistant.

Finally, it was demonstrated that native α-La possesses several classes of fatty acid binding site [52].

Some multimeric human α-La derivative induces an increase of Ca^{2+} level and acts as an apoptosis-inducing agent [53,54]. In fact, it has been shown that multimeric α-La can bind to the cell surface, enter in the cytoplasm and accumulate in the nuclei. Further, the direct interaction of α-La with mitochondria lead to cytochrome c release, which, in turn, start the caspase cascade which is involved in apoptosis [55].

In particular, it has been shown that in human milk the α-lactalbumin-oleic acid complex, called HAMLET (Human α-lactalbumin Made Lethal to Tumor Cells), is capable of inducing selective apoptosis towards cancer cells, but leaves fully differentiated cells unaffected. The in vivo effects of HAMLET have been investigated in patients and in tumor cell lines, and the results indicated that HAMLET limits the progression of human glioblastomas in a xenograft model and removes skin papillomas in patients. In tumor cells, HAMLET enters the cytoplasm and then enters the nuclei where it accumulates. Here, HAMLET binds to histones and disrupts chromatin organization [56].

Furthermore, it has been shown that the acid pH in the stomach of the breast-fed child can promote the formation of HAMLET, which contribute to the protective effect of breastfeeding against childhood tumors [56].

HAMLET can be produced starting from a purified native α-lactalbumin which is initially partially unfolded by ethylenediaminetetraacetic acid (EDTA), a chelating agent able to remove the calcium ion. The protein adopts the apo-conformation that exposes a new fatty acid binding site, which has high specificity for oleic acid.

Subsequently, the protein is subjected to an ion-exchange chromatography on a matrix conditioned with oleic acid, which binds to α-La and stabilize the altered protein conformation. The complex eluted from the column in presence of high salt solution is the HAMLET complex [56].

3.2.3. Lysozyme

In donkey milk, lysozyme has been purified and characterized by a reversed-phase HPLC, followed by a 15% polyacrylamide gel electrophoresis under denaturing conditions (15% SDS-PAGE). Two-dimensional electrophoresis revealed the presence of one spot corresponding to the lysozyme, with a Mr of 14.5 kDa and a pI of 9.40 [37,38]. However, in the literature are reported the presence of two isoforms of donkey milk lysozyme: lysozyme A (Mr 14.631 kDa) and lysozyme B (Mr 14.646 kDa), which differ for an oxidized methionine at position 79. This derives, most likely, from a post-translational modification [44,45]. Donkey milk contains 1.0 g/L of lysozyme (higher with respect to the lysozyme content of human milk), while in goat and bovine milk this protein is virtually absent. Lysozyme belongs to the class of hydrolases, and specifically is a glycosidase. It has a structure consisting of two domains: one domain is essentially composed of α-helices, and the other consists of an anti-parallel β-sheet and two α-helices. The three-dimensional configuration of the molecule is maintained by the presence of three disulphide bridges: two located in the alpha-helix domain, and one in the β-sheet. In its inner part there are few polar residues.

Under the bioactive aspect, this enzyme exerts a bactericidal action, as it disrupts the cell wall of the microorganisms; its hydrolytic reaction occurs only on the *O*-glycosyl compounds. This results in the cleavage of the glycoside bond between the carbon 1 of the glucosamine and the carbon 4 of the muramic acid, thereby causing the broken of bacterial cell wall.

Various experiments have shown that due to the high content of lysozyme in donkey milk, raw milk of this species, different to raw cow milk, does not change its organoleptic and microbiological characteristics over time. This suggests a long shelf life of donkey milk which could be stored longer with respect to raw cow milk [57]. In another study on the antimicrobial properties of donkey milk, Šarić and co-workers found a positive correlation between the high content of lysozyme and the antibacterial activity against *Listeria monocytogenes* and *Staphylococcus aureus* [58].

Furthermore, thanks to its bactericidal action, lysozyme may help to prevent intestinal infections in infants, thus facilitating a proper digestion and normal absorption of nutrients.

In addition, it has been shown that this enzyme has other physiological functions: anti-inflammatory activity, immunoregulatory activity, and antitumor activity. In particular, studies have shown that lysozyme inhibits angiogenesis and has antitumor activity. Recent in vitro studies have noted an anti-proliferative and antitumor activity of donkey milk against the adenocarcinoma human alveolar basal epithelial cells (A549) through the stimulation of IL-2, IFN-g, IL-6, TNF-a, and IL-1b cytokines production, the modification of the cell cycle, and the induction of the apoptosis process [59]. Furthermore, it has been shown that equine lysozyme, which, as discussed before, is structurally homologous to α-lactalbumin, can form complexes with oleic acid called ELOA (equine lysozyme and oleic acid). This, in analogy to the HAMLET complex, may show cytotoxic activity. Recently, it has been shown that ELOA also displays bactericidal activity against pneumococci—in particular these complex binds to *Streptococcus pneumoniae*, causing perturbations of the plasma membrane, such as depolarization and rupture, and consequently calcium enters the cells. As discussed before, the increase of intracellular calcium induces apoptosis. Furthermore, analogously to HAMLET,

ELOA-induced apoptosis is accompanied by DNA fragmentation into high molecular weight fragments [60]. Equine lysozyme also possesses the high-affinity calcium-binding site of α-lactalbumin, whose amino acid sequence is highly conserved [56].

3.2.4. Lactoferrin

Lactoferrin, also called lactotransferrin, is an 80.0 kDa glycoprotein belonging to the transferrin family, is an iron-chelating glycoprotein, and has a structure composed of two homologous domains, each binding a ferric ion (Fe^{3+}) and a carbonate anion. It is a protein that shows several functions, including regulation of iron homoeostasis, cell growth and differentiation, defense against infectious agents, anti-inflammatory and cancer protection, and ultimately a trophic activity on the intestinal mucosa [61]. Lactoferrin is present mainly in milk, and also in small quantities in exocrine fluids such as saliva, tears, bile, seminal fluid, and pancreatic juice. The plasma contains a low concentration of lactoferrin, but during inflammatory reactions, the neutrophil granulocytes release this protein by increasing the concentration of the plasma pool. Lactoferrin concentration is quite high in human milk (1.0 mg/mL) compared with bovine milk (0.02–0.2 g/L), sheep (0.14 g/mL), and goat (0.02–0.40 g/L). However, in all species, the highest lactoferrin concentration is found in the colostrum (in the human colostrum is about 7.0 g/L) and may increase in the case of breast infection. In mare milk, lactoferrin concentration is about 0.10 g/L, while in colostrum ranges from 1.5 to 5.0 g/L. In donkey milk was found a lactoferrin concentration of 0.08 g/L, close to that of mare milk, cow, and goat, but lower than the average concentration found in human milk [38]. Lactoferrin shows two different mechanisms of antimicrobial activity. The first mechanism has a bacteriostatic effect, because, thanks to its high affinity with iron, it can subtract this metal to iron-dependent bacteria, which are thus deprived of essential nutritious growth. Thanks to its bacteriostatic properties, due to the ability to bind iron, lactoferrin is capable of delaying growth to a wide variability of microorganisms, including a wide range of gram-positive and gram-negative bacteria and some type of yeasts. However, the bacteriostatic effect is often temporary because some gram-negative bacteria became able to adapt themselves to the restrictive conditions of iron through the synthesis of low molecular weight chelating agents (siderophores) that are able to remove iron from lactoferrin. The second antimicrobial activity is due to a direct-action mechanism of lactoferrin that is capable of damaging the cell wall of gram-negative bacteria, binding to lipopolysaccharide A, to the porins, and other surface molecules of the cell wall of some microorganisms. Lactoferrin has an important anti-inflammatory role associated with the microbial challenge. It has been shown through animal studies that the administration of lactoferrin protects against gastritis acting against *Helicobacter pylori* [62]. This protective effect against infections seems to be due to the fact that lactoferrin promotes the inhibition of several pro-inflammatory cytokines such as tumor necrosis factor alpha (TNFa), interleukin-1β (IL-1β) and IL-6. Lactoferrin is upregulated in several inflammatory disorders such as neurodegenerative disease, arthritis, allergic skin, inflammatory bowel disease, and lung disorders [61]. Furthermore, lactoferrin, besides the many biological roles already known, stimulates the proliferation and differentiation of osteoblasts as shown by Cornish and co-workers [63].

From the digestion of lactoferrin derive some peptides with antimicrobial activity against pathogens, namely LF1-11, lactoferrampin, and lactoferricin. All of these derive from the N-terminal domain of lactoferrin and are conserved in lactoferrin of most species. The most important peptide is lactoferricin which exerts its antibacterial activity against several bacteria, viruses, fungal pathogens, and protozoa. Furthermore, this peptide displayed other activities, such as inhibition of tumor metastasis in mice and induction of apoptosis in THP-1 human monocytic leukemic cells. [64].

Comparing the antimicrobial activities of lactoferricin from human, bovine, murine, and caprine, it was shown that bovine lactoferricin was the most active. In fact, its minimal inhibitory concentration (MIC) against some *Escherichia coli* strains was about 30 µg/Ml, whereas in the lactoferricin derived from human lactoferrin, IT was about 4 times higher. The higher antimicrobial activity of bovine lactoferricin seems to be due to the presence of high amounts of net positive charges and hydrophobic residues [64]. Another group of peptides which derives from lactoferrin digestion includes lactoferroxins. These are classified as opioid peptides, since they show effects similar to that of opiates [65]. Opioid peptides can be also obtained from the digestion of caseins (casomorphines).

3.2.5. Lactoperoxidase

Lactoperoxidase (LPO) is an enzyme predominantly secreted by the mammal gland, but it can be found also in other glandular secretions. This enzyme is part of the peroxidase family and is a glycoprotein consisting of 608 amino acids with a molecular mass of 78.0 kDa. Its secondary structure consists of α-helices and two short-anti-parallel β-sheets which form, together, a spheroidal structure with a heme group in the center linked covalently. To the enzyme is also linked with a calcium ion, important for the maintaining of the protein structural integrity [66]. The LPO system catalyzes the oxidation of several substrates, by using hydrogen peroxide (H_2O_2) as shown in the following reaction:

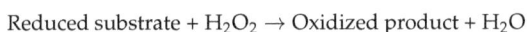

$$\text{Reduced substrate} + H_2O_2 \rightarrow \text{Oxidized product} + H_2O$$

Reduced substrates include thiocyanate (SCN^-) and iodide ions (I^-), whereas hydrogen peroxide can derive from the reaction of glucose with oxygen catalyzed by the enzyme glucose oxidase. Once formed, the oxidized products have a potent bactericidal activity against bacteria, virus, parasite, fungi, and mycoplasma [67]. LPO system has a bacteriostatic effect also on *Listeria monocytogenes*, therefore, it should be used for controlling the development of this bacterium in the raw milk at refrigeration temperatures. Furthermore, some authors co-administered lactoferrin and LPO to mice infected with influenza virus in order to attenuate pneumonia [68]. Thanks to its bacteriostatic activity, LPO is widely used in the preservation of food.

Donkey milk LPO enzymatic activity was 4.83 ± 0.35 mU/mL and its amount 0.11 ± 0.027 mg/L [38]. This value resulted 100 times lower with respect to the bovine milk LPO (0.03–0.1 g/L), but very similar to that found in human milk (0.77 ± 0.38 mg/L) [67,69].

The three above-mentioned protective factors (lysozyme, LPO, and lactoferrin) may have additive or synergic effects, as reported by some authors [70,71], and, as known, all of them contribute to the protection of the newborn animal. It is interesting to notice that these antimicrobial factors are almost the same in different species, but their amount and importance can vary widely. In fact, the content of lysozyme in donkey and human milk is considerably higher than that of bovine milk, whereas lactoperoxidase is present in small amounts in donkey and human milk, but is high in bovine milk (Table 8).

Table 8. Amount of the three main antimicrobial factors in human, bovine, and donkey milk.

Milk	Lysozyme (g/L)	Lactoperoxidase (mg/L)	Lactoferrin (g/L)
Human	0.12	0.77	0.3–4.2
Donkey	1.0	0.11	0.08
Bovine	Trace	30–100	0.10

Adapted from [38].

Furthermore, it has been demonstrated that the immunoglobulins found in the milk exert a synergistic effect on the activity of these non-specific antimicrobial factors, as demonstrated by Tenuovo and co-workers [70] in the case of the LPO system, whose antimicrobial activity against *Streptococcus mutans* increases considerably if the system is incubated with secretory IgA. The antimicrobial activity enhancement seems to be due to a binding between LPO and immunoglobulins (IgA). The result of this interaction is a stabilization of the enzymatic activity of lactoperoxidase.

3.2.6. β-Casein Fragments

Bioactive peptides derive from caseins and they possess multifunctional properties. In fact, they initially act as physiological modulators of metabolism during intestinal digestion of food and when absorbed, they can act on the various target organs of the organism. In particular, bioactive peptides play several roles in the cardiovascular, digestive, nervous, and immune systems [72].

Cunsolo and co-workers [45] found the presence of β-casein fragments in the donkey milk whey protein fraction, which can derive from the partial digestion of β-casein by endogenous proteases. In particular, these authors found that the donkey's β-casein fragment 199–226, with molecular mass of 3043.0 Da, share 40% sequence homology with the human β-casein fragment 184–210, which shows antimicrobial activity. Furthermore, the N-terminal sequence of the donkey β-casein fragment 199–226, has homology with the bovine β-casein fragment 193–198, which has anti-hypertensive properties [73] (Figure 4).

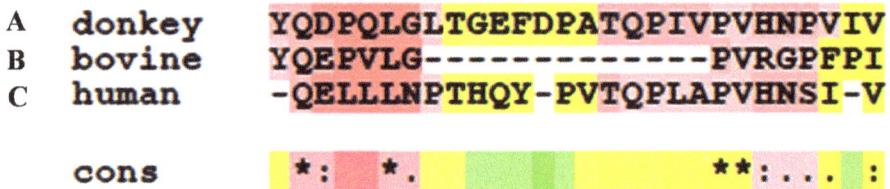

Figure 4. Alignment of fragments derived from the β-casein digestion with anti-hypertensive properties: (**A**): donkey's β-casein fragment 199–226; (**B**): human β-casein fragment 184–210; (**C**): bovine β-casein fragment 193–198 [45]. Sequence alignment was performed by T-COFFEE, Version_11.00 [74].

However, the function and the impact on human health of the bioactive peptides derived from donkey milk casein is not well elucidated yet. Therefore, for the time being, it is possible only to speculate on the possible function of these peptides based on sequence similarities with the peptides derived from bovine milk [45]. Some properties of bioactive peptides derived from bovine milk casein are reported below.

Some authors showed that some peptides derived from αs_1-casein and from β-caseins known as casokinins (particularly fragments 177–183 and 193–202) inhibit the in vitro activity of angiotensin-converting-enzyme (ACE) [75], thus acting as anti-hypertensive agents. A similar result was obtained by other authors, which showed that some peptides derived from milk whey and casein proteins possess ACE-inhibitory activity, in particular the peptides identified were from αs_1-casein (fragments 142–147, 157–164 and 194–199) and β-caseins (fragments 108–113, 177–183 and 193–198) [76].

The in vitro enzymatic hydrolysis of β-casein produces several peptides with opioid-like activity and characterized by a common N-terminal sequence Tyr-Gly-Gly-Phe. The presence of a Tyr residue at the N-terminus is important for the opioid activity. The most important opioid peptides are the β-casomorphins, which are fragments of β-caseins between the 60th and 70th residues [72]. The presence of these bioactive peptide has effect on central nervous system (analgesic and sedative action), and on the endocrine system.

Furthermore, the bioactive peptides derived from caseins also show also immunomodulatory activity (stimulation of the immune system) and antimicrobial activity (inhibition of pathogens). It has been proven that the digestion of human and bovine milk caseins releases peptides with immunostimulating activity. Parker and co-workers [77] demonstrated in particular the immunomodulatory activity of the fragment 54–50 of human β-casein (hexapeptide Val-Glu-Pro-Ile-Pro-Tyr).

4. Conclusions

Donkey milk could be considered a valid replacement for human milk for clinical tolerability, palatability, and nutritional adequacy for children affected by CMPA. It could also furnish additional physiological functions, such as providing antibacterial substances, digestive activity molecules, growth factors, and hormones. In children's nutrition, the use of natural milk, rather than a formula during the cow's milk free diet period, should be encouraged. In this context, it is very important that the nutraceutical role is discovered for donkey milk, because of its content of functional proteins and peptides that have immunological-like properties and are able to stimulate the functional recovery and development of the neonatal intestine. In particular, in the whey protein fraction, it is worth mentioning the presence of a good amount of α-lactalbumin which, besides being implicated in lactose synthesis, also induces apoptosis in cancer cells when complexed with oleic acid (HAMLET) as demonstrated by in vivo and in vitro experiments. A high amount of lysozyme which shows a bactericidal action against several pathogens and, similarly to α-lactalbumin, can form complex with oleic acid with bactericidal action called ELOA. Donkey milk contains a discrete amount of lactoferrin which display several positive effects on human health such as iron homeostasis, antimicrobial activity, cellular growth differentiation, anti-inflammatory activity, protection against cancer development, and metastasis. Furthermore, once digested in the stomach, lactoferrin is fragmented into small peptides with a more potent action against several bacteria, viruses, fungal pathogens and protozoa, with respect to the full-length lactoferrin from which they derive. Finally, in donkey milk whey protein fraction, the presence of small peptides derived from β-casein, whose sequence was found to be similar to that of peptides derived from bovine β-caseins, was demonstrated. The function of these peptides derived from donkey milk β-casein is not well elucidated yet, but it can be supposed that they may have roles on cardiovascular, digestive, nervous, and immune systems as described for the bovine counterpart.

The structural similarities between human and donkey milk proteins could contribute to explain the results of clinical studies performed by using this milk [22,26,78] indicating that donkey milk may be a valid substitute of cow milk for feeding allergic children. Furthermore, the nutraceutical and functional properties determined in donkey milk could be considered valid reasons for the production of starting formulas based on donkey milk.

However, in the use of donkey milk for infant nutrition, it should also be taken into account the low lipid content, and consequently its low energetic value, which could limit the use of this kind of milk for feeding children affected by CMPA [79]. For this purpose, more clinical studies designed to evaluate the nutritional efficiency of donkey milk in the first years of life should be encouraged, since until now the case studies reported in literature are still too few.

Conflicts of Interest: The authors declare no conflict of interest.

References

1. Claeys, W.L.; Verraes, C.; Cardoen, S.; De Block, J.; Huyghebaert, A.; Raes, K.; Dewettinck, K.; Herman, L. Consumption of raw or heated milk from different species: An evaluation of the nutritional and potential health benefits. *Food Control* **2014**, *42*, 188–201. [CrossRef]
2. Thompson, A.; Boland, M.; Singh, A. Milk proteins from expression to food. In *Food Sciences and Technology*; Thompson, A., Boland, M., Singh, A., Eds.; Elsevier: Burlington, MA, USA, 2009; pp. 8–56.
3. Ragona, G.; Corrias, F.; Benedetti, M.; Paladini, M.; Salari, F.; Altomonte, L.; Martini, M. Amiata Donkey Milk Chain: Animal Health Evaluation and Milk Quality. *Ital. J. Food Saf.* **2016**, *5*, 5951. [CrossRef] [PubMed]
4. Fox, P.F.; McSweeney, P.L.H. *Dairy Chemistry and Biochemistry*; Blackie Academic and Professional: London, UK, 1998; pp. 67–71.
5. Donovan, S.M.; Monaco, M.H.; Bleck, G.T.; Cook, J.B.; Noble, M.S.; Hurley, W.L.; Wheeler, M.B. Transgenic Over-Expression of Bovine α-Lactalbumin and Human Insulin-Like Growth Factor-I in Porcine Mammary Gland. *J. Dairy Sci.* **2001**, *84*, E216–E222. [CrossRef]

6. Haug, A.; Høstmark, A.T.; Harstad, O.M. Bovine milk in human nutrition—A review. *Lipids Health Dis.* **2007**, *6*, 25. [CrossRef] [PubMed]

7. Rafiq, S.; Huma, N.; Pasha, I.; Sameen, A.; Mukhtar, O.; Khan, M.I. Chemical Composition, Nitrogen Fractions and Amino Acids Profile of Milk from Different Animal Species. *Asian Australas. J. Anim. Sci.* **2016**, *29*, 1022–1028. [CrossRef] [PubMed]

8. Holt, C.; Jenness, R. Interrelationships of constituents and partition of salts in milk samples from eight species. *Comp. Biochem. Physiol. A Comp. Physiol.* **1984**, *77*, 275–282. [CrossRef]

9. Anderson, R.R. Comparison of trace elements in milk of four species. *J. Dairy Sci.* **1992**, *75*, 3050–3055. [CrossRef]

10. Uniacke-Lowe, T.; Fox, P.F. Equid Milk: Chemistry, Biochemistry and Processing. In *Food Biochemistry and Food Processing*, 2nd ed.; Simpson, B.K., Ed.; John Wiley & Sons, Inc.: Oxford, UK, 2012; pp. 491–528.

11. Buttriss, J. *Adverse Reaction to Food. The Report of a British Nutrition Foundation Task Force*; Buttriss, J., Ed.; Blackwell Science: Oxford, UK, 2002.

12. Hochwallner, H.; Schulmeister, U.; Swoboda, I.; Spitzauer, S.; Valenta, R. Cow's milk allergy: From allergens to new forms of diagnosis, therapy and prevention. *Methods* **2014**, *66*, 22–33. [CrossRef] [PubMed]

13. Pessler, F.; Nejat, M. Anaphylactic reaction to goat's milk in a cow's milk—Allergic infant. *Pediatr. Allergy Immunol.* **2004**, *15*, 183–185. [CrossRef] [PubMed]

14. Ghosh, J.; Malhotra, G.S.; Mathur, B.N. Hypersensitivity of human subjects to bovine milk proteins: A review. *Indian J. Dairy Sci.* **1989**, *42*, 744–749.

15. Giner, M.T.; Vasquez, M.; Dominiguez, O. Specific oral desensitization in children with IgE-mediated cow's milk allergy. Evolution in one year. *Eur. J. Pediatr.* **2012**, *171*, 1389–1395. [CrossRef]

16. Sánchez-García, S.; del Río, P.R.; Escudero, C.; García-Fernández, C. Efficacy of Oral Immunotherapy Protocol for Specific Oral Tolerance Induction in Children with Cow's Milk Allergy. *IMAJ* **2012**, *14*, 43–47. [PubMed]

17. Venter, C. Cow's milk protein allergy and other food hypersensitivities in infants. *J. Fam. Health Care* **2009**, *19*, 128–134. [PubMed]

18. Wang, J. Management of the Patient with Multiple Food Allergies. *Curr. Allergy Asthma Rep.* **2010**, *10*, 271–277. [CrossRef] [PubMed]

19. Cunsolo, V.; Muccilli, V.; Fasoli, E.; Saletti, R.; Righetti, R.G.; Foti, S. Poppea's bath liquor: The secret proteome of she-donkey's milk. *J. Proteom.* **2011**, *74*, 2083–2099. [CrossRef] [PubMed]

20. Jirillo, F.; Jirillo, E.; Magrone, T. Donkey's and goat's milk consumption and benefits to human health with special reference to the inflammatory status. *Curr. Pharm. Des.* **2010**, *16*, 859–863. [CrossRef] [PubMed]

21. Polidori, P.; Ariani, A.; Vincenzetti, S. Use of Donkey Milk in Cases of Cow's Milk Protein Allergies. *Int. J. Child Health Nutr.* **2015**, *4*, 174–179. [CrossRef]

22. Iacono, G.; Carroccio, A.; Cavataio, F.; Montalto, G.; Soresi, M.; Balsamo, V. Use of ass's milk in multiple food allergy. *J. Pediatr. Gastroenterol. Nutr.* **1992**, *14*, 177–181. [CrossRef] [PubMed]

23. Polidori, P.; Beghelli, D.; Mariani, P.; Vincenzetti, S. Donkey milk production: State of the art. *Ital. J. Anim. Sci.* **2009**, *8*, 677–683. [CrossRef]

24. Salimei, E.; Fantuz, F.; Coppola, R.; Chiofalo, B.; Polidori, P.; Varisco, G. Composition and characteristics of ass's milk. *Anim. Res.* **2004**, *53*, 67–78. [CrossRef]

25. Malacarne, M.; Martuzzi, F.; Summer, A.; Mariani, P. Protein and fat composition of mare's milk: Some nutritional remarks with reference to human and cow's milk. *Int. Dairy J.* **2002**, *12*, 869–877. [CrossRef]

26. Carroccio, A.; Cavataio, F.; Montalto, G.; D'Amico, D.; Alabrese, L.; Iacono, G. Intolerance to hydrolised cow's milk proteins in infants: Clinical characteristics and dietary treatment. *Clin. Exp. Allergy* **2000**, *30*, 1597–1603. [CrossRef] [PubMed]

27. Monti, G.; Viola, S.; Baro, C.; Cresi, F.; Tovo, P.A.; Moro, G.; Ferrero, M.P.; Conti, A.; Bertino, E. Tolerability of donkey's milk in 92 highly-problematic cow's milk allergic children. *J. Biol. Regul. Homeost. Agents* **2012**, *26*, 75–82. [PubMed]

28. Chiofalo, B.; Salimei, E.; Chiofalo, L. Acidi grassi nel latte d'asina: Proprietà bio-nutrizionali ed extranutrizionali. *Large Anim. Rev.* **2003**, *6*, 21–26.

29. Martemucci, G.; D'Alessandro, A.G. Fat content, energy value and fatty acid profile of donkey milk during lactation and implications for human nutrition. *Lipids Health Dis.* **2012**, *11*, 113. [CrossRef] [PubMed]

30. Vincenzetti, S.; Pucciarelli, S.; Nucci, C.; Polzonetti, V.; Cammertoni, N.; Polidori, P. Profile of nucleosides and nucleotides in donkey's milk. *Nucleosides Nucleotides Nucleic Acids* **2014**, *33*, 656–667. [CrossRef] [PubMed]

31. Birlouez-Aragon, I. Effect of lactose hydrolysis on calcium absorption during duodenal milk perfusion. *Reprod. Nutr. Dev.* **1988**, *28*, 1465–1472. [CrossRef] [PubMed]

32. Griessen, M.; Cochet, B.; Infante, F.; Jung, A.; Bartholdi, P.; Donath, A.; Loizeau, E.; Courvoisier, B. Calcium absorption from milk in lactase-deficient subjects. *Am. J. Clin. Nutr.* **1989**, *49*, 377–384. [PubMed]

33. Chiavari, C.; Coloretti, F.; Nanni, M.; Sorrentino, E.; Grazia, L. Use of donkey's milk for a fermented beverage with lactobacilli. *Lait* **2005**, *85*, 481–490. [CrossRef]

34. Vincenzetti, S.; Polidori, P.; Vita, A. Nutritional characteristics of donkey's milk protein fraction. In *Dietary Protein Research Trends*; Ling, J.R., Ed.; Nova Science Publisher Inc.: New York, NY, USA, 2007; pp. 207–225.

35. Vita, D.; Passalacqua, G.; Di Pasquale, G.; Caminiti, L.; Crisafulli, G.; Rulli, I.; Pajno, G.B. Ass's milk in children with atopic dermatitis and cow's milk allergy: Crossover comparison with goat's milk. *Pediatr. Allergy Immunol.* **2007**, *18*, 594–598. [CrossRef] [PubMed]

36. Martin, P.; Grosclaude, F. Improvement of milk protein-quality by gene technology. *Livest. Prod. Sci.* **1993**, *35*, 95–115. [CrossRef]

37. Vincenzetti, S.; Polidori, P.; Mariani, P.; Cammertoni, N.; Fantuz, F.; Vita, A. Donkey milk protein fractions characterization. *Food Chem.* **2008**, *10*, 640–649. [CrossRef]

38. Vincenzetti, S.; Amici, A.; Pucciarelli, S.; Vita, A.; Micozzi, D.; Carpi, F.M.; Polzonetti, V.; Natalini, P.; Polidori, P. A Proteomic Study on Donkey Milk. *Biochem. Anal. Biochem.* **2012**, *1*, 109. [CrossRef]

39. Greppi, G.F.; Roncada, P. La componente proteica del latte caprino. In *L'alimentazione della Capra da Latte*; Pulina, G., Ed.; Avenue Media Publisher: Bologna, Italy, 2005; pp. 71–99.

40. Bertino, E.; Gastaldi, D.; Monti, G.; Baro, C.; Fortunato, D.; Perono Garoffo, L.; Coscia, A.; Fabris, C.; Mussap, M.; Conti, A. Detailed proteomic analysis on DM: Insight into its hypoallergenicity. *Front. Biosci.* **2010**, *2*, 526–536. [CrossRef]

41. Chianese, L.; Calabrese, M.G.; Ferranti, P.; Mauriello, R.; Garro, G.; De Simone, C.; Quarto, M.; Addeo, F.; Cosenza, G.; Ramunno, L. Proteomic characterization of donkey milk "caseome". *J. Chromatogr. A* **2010**, *1217*, 4834–4840. [CrossRef] [PubMed]

42. Godovac-Zimmermann, J.; Conti, A.; James, L.; Napolitano, L. Microanalysis of the amino-acid sequence of monomeric beta-lactoglobulin I from donkey (*Equus asinus*) milk. The primary structure and its homology with a superfamily of hydrophobic molecule transporters. *Biol. Chem. Hoppe-Seyler* **1988**, *369*, 171–179. [CrossRef] [PubMed]

43. Godovac-Zimmermann, J.; Conti, A.; Sheil, M.; Napolitano, L. Covalent structure of the minor monomeric beta-lactoglobulin II component from donkey milk. *Biol. Chem. Hoppe-Seyler* **1990**, *371*, 871–879. [CrossRef] [PubMed]

44. Herrouin, M.; Molle, D.; Fauquant, J.; Ballestra, F.; Maubois, J.L.; Leonil, J. New Genetic Variants Identified in Donkey's Milk Whey Proteins. *J. Protein Chem.* **2000**, *19*, 105–115. [CrossRef] [PubMed]

45. Cunsolo, V.; Saletti, R.; Muccilli, V.; Foti, S. Characterization of the protein profile of donkey's milk whey fraction. *J. Mass Spectrom.* **2007**, *42*, 1162–1174. [CrossRef] [PubMed]

46. Kontopidis, G.; Holt, C.; Sawyer, L. Invited review: Beta-lactoglobulin: Binding properties, structure, and function. *J. Dairy Sci.* **2004**, *87*, 785–796. [CrossRef]

47. Liang, L.; Tajmir-Riahi, H.A.; Subirade, M. Interaction of beta-lactoglobulin with resveratrol and its biological implications. *Biomacromolecules* **2008**, *9*, 50–56. [CrossRef] [PubMed]

48. Liang, L.; Subirade, M. Beta-Lactoglobulin/Folic Acid Complexes: Formation, Characterization, and Biological Implication. *J. Phys. Chem. B* **2010**, *114*, 6707–6712. [CrossRef] [PubMed]

49. Permyakova, E.A.; Berliner, L.J. α-Lactalbumin: Structure and function. *FEBS Lett.* **2000**, *473*, 269–274. [CrossRef]

50. Pelligrini, A.; Thomas, U.; Bramaz, N.; Hunziker, P.; von Fellenberg, R. Isolation and identification of three bactericidal domains in the bovine alpha-lactalbumin molecule. *Biochim. Biophys. Acta* **1999**, *1426*, 439–448. [CrossRef]

51. Hakansson, A.; Svensson, M.; Mossberg, A.K.; Sabharwal, H.; Linse, S.; Lazou, I.; Lonnerdal, B.; Svanborg, C. A folding variant of alpha-lactalbumin with bactericidal activity against *Streptococcus pneumoniae*. *Mol. Microbiol.* **2000**, *35*, 589–600. [CrossRef] [PubMed]

52. Cawthern, K.M.; Narayan, M.; Chaudhuri, D.; Permyakov, E.A.; Berliner, L.J. Interactions of α-Lactalbumin with Fatty Acids and Spin Label Analogs. *J. Biol. Chem.* **1997**, *272*, 30812–30816. [CrossRef] [PubMed]

53. Hakansson, A.; Zhivotovsky, B.; Orrenius, S.; Sabharwal, H.; Svanborg, C. Apoptosis induced by a human milk protein. *Proc. Natl. Acad. Sci. USA* **1995**, *92*, 8064–8068. [CrossRef] [PubMed]

54. Svensson, M.; Sabharwal, H.; Hakansson, A.; Mossberg, A.K.; Lipniunas, P.; Leffler, H.; Svanborg, C.; Linse, S. Molecular Characterization of α–Lactalbumin Folding Variants That Induce Apoptosis in Tumor Cells. *J. Biol. Chem.* **1999**, *274*, 6388–6396. [CrossRef] [PubMed]

55. Köhler, C.; Hakansson, A.; Svanborg, C.; Orrenius, S.; Zhivotovsky, B. Protease activation in apoptosis induced by MAL. *Exp. Cell Res.* **1999**, *249*, 260–268. [CrossRef] [PubMed]

56. Mossberg, A.K.; Hun Mok, K.; Morozova-Roche, L.A.; Svanborg, C. Structure and function of human α-lactalbumin made lethal to tumor cells (HAMLET)-type complexes. *FEBS J.* **2010**, *277*, 4614–4625. [CrossRef] [PubMed]

57. Zhang, X.Y.; Zhao, L.; Jiang, L.; Dong, M.L.; Ren, F.Z. The antimicrobial activity of donkey milk and its microflora changes during storage. *Food Control* **2008**, *19*, 1191–1195. [CrossRef]

58. Šarić, L.C.; Šarić, B.M.; Kravić, S.T.; Plavšić, D.V.; Milovanović, I.L.; Gubić, J.M.; Nedeljković, N.M. Antibacterial activity of domestic Balkan donkey milk toward *Listeria monocytogenes* and *Staphylococcus aureus*. *Food Feed Res.* **2014**, *41*, 47–54. [CrossRef]

59. Mao, X.; Gu, J.; Sun, Y.; Xu, S.; Zhang, X.; Yang, H.; Ren, F. Anti-proliferative and anti-tumour effect of active components in donkey milk on A549 human lung cancer cells. *Int. Dairy J.* **2009**, *19*, 703–708. [CrossRef]

60. Clementi, E.A.; Wilhelm, K.R.; Schleicher, J.; Morozova-Roche, L.A.; Hakansson, A.P. A complex of equine lysozyme and oleic acid with bactericidal activity against *Streptococcus pneumoniae*. *PLoS ONE* **2013**, *8*, e80649. [CrossRef] [PubMed]

61. Ward, P.P.; Paz, E.; Conneely, O.M. Multifunctional roles of lactoferrin: A critical overview. *Cell. Mol. Life Sci.* **2005**, *62*, 2540–2548. [CrossRef] [PubMed]

62. Dial, E.J.; Lichtenberger, L.M. Effect of lactoferrin on *Helicobacter felis* induced gastritis. *Biochem. Cell Biol.* **2002**, *80*, 113–117. [CrossRef] [PubMed]

63. Cornish, J.; Palmano, K.; Callon, K.E.; Watson, M.; Lin, J.M.; Valenti, P.; Naot, D.; Grey, A.B.; Reid, I.R. Lactoferrin and bone; structure-activity relationships. *Biochem. Cell. Biol.* **2006**, *84*, 297–302. [CrossRef] [PubMed]

64. Sinha, M.; Kaushik, S.; Kaur, P.; Sharma, S.; Singh, T.P. Antimicrobial Lactoferrin Peptides: The Hidden Players in the Protective Function of a Multifunctional Protein. *Int. J. Pept.* **2013**, *2013*, 390230. [CrossRef] [PubMed]

65. Jenssen, H. Antimicrobial activity of lactoferrin and lactoferrin derived peptides. In *Dietary Protein Research Trends*; Ling, J.R., Ed.; Nova Science Publisher Inc.: New York, NY, USA, 2007; pp. 1–62.

66. Tenovuo, J.O. The peroxidase system in human secretions. In *The Lactoperoxidase System: Chemistry and Biological Significance*; Pruitt, K.M., Tenovuo, J.O., Eds.; Marcel Dekker: New York, NY, USA, 1985; pp. 101–122.

67. Tanaka, T. Antimicrobial activity of lactoferrin and lactoperoxidase in milk. In *Dietary Protein Research Trends*; Ling, J.R., Ed.; Nova Science Publisher Inc.: New York, NY, USA, 2007; pp. 101–115.

68. Shin, K.; Wakabayashi, H.; Yamauchi, K.; Teraguchi, S.; Tamura, Y.; Kurokawa, M.; Shiraki, K. Effects of orally administered bovine lactoferrin and lactoperoxidase on influenza virus infection in mice. *J. Med. Microbiol.* **2005**, *54*, 717–723. [CrossRef] [PubMed]

69. Shin, K.; Hayasawa, H.; Lönnerdal, B. Purification and quantification of lactoperoxidase in human milk with use of immunoadsorbent with antibodies against recombinant human lactoperoxidase. *Am. J. Clin. Nutr.* **2001**, *73*, 984–989. [PubMed]

70. Tenovuo, J.; Moldoveanu, Z.; Mestecky, J.; Pruitt, K.M.; Rahemtulla, B.M. Interaction of specific and innate factors of immunity: IgA enhances the antimicrobial effect of the lactoperoxidase system against *Streptococcus mutans*. *J. Immunol.* **1982**, *128*, 726–731. [PubMed]

71. Arnold, R.; Russell, J.E.; Devine, S.M.; Adamson, M.; Pruitt, K.M. Antimicrobial activity of the secretory innate defence factors lactoferrin, lactoperoxidase and lysozyme. In *Cardiology Today*; Guggenheim, B., Ed.; S. Karger: Basel, Switzerland, 1984; pp. 75–88.

72. Silva, S.V.; Malcata, F.X. Caseins as a source of bioactive peptides. *Int. Dairy J.* **2005**, *15*, 1–15. [CrossRef]

73. Minervini, F.; Algaron, F.; Rizzello, C.G.; Fox, P.F.; Monnet, V.; Gobbetti, M. Angiotensin I-converting-enzyme-inhibitory and antibacterial peptides from *Lactobacillus helveticus* PR4 proteinase-hydrolyzed caseins of milk from six species. *Appl. Environ. Microbiol.* **2003**, *69*, 5297–5305. [CrossRef] [PubMed]

Beverages **2017**, *3*, 34

74. Notredame, C.; Higgins, D.G.; Heringa, J. T-Coffee: A novel method for fast and accurate multiple sequence alignment. *J. Mol. Biol.* **2000**, *302*, 205–217. [CrossRef] [PubMed]

75. Maruyama, S.; Suzuki, H. A peptide inhibitor of angiotensin I-converting enzyme in the tryptic hydrolysate of casein. *Agric. Biol. Chem.* **1982**, *46*, 1393–1394. [CrossRef]

76. Pihlanto-Leppala, A.; Rokka, T.; Korhonen, H. Angiotensin I converting enzyme inhibitory peptides from bovine milk proteins. *Int. Dairy J.* **1998**, *8*, 325–331. [CrossRef]

77. Parker, F.; Migliore-Samour, D.; Floch, F.; Zerial, A.; Werner, G.H.; Jollès, J.; Casaretto, M.; Zahn, H.; Jollès, P. Immunostimulating hexapeptide from human casein: Amino acid sequence, synthesis and biological properties. *Eur. J. Biochem.* **1984**, *145*, 677–682. [CrossRef] [PubMed]

78. Tesse, R.; Paglialunga, C.; Braccio, S.; Armenio, L. Adequacy and tolerance to ass's milk in an Italian cohort of children with cow's milk allergy. *Ital. J. Pediatr.* **2009**, *35*, 19. [CrossRef] [PubMed]

79. Giovannini, M.; D'Auria, E.; Caffarelli, C.; Verduci, E.; Barberi, S.; Indinnimeo, L.; Iacono, I.D.; Martelli, A.; Riva, E.; Bernardini, R. Nutritional management and follow up of infants and children with food allergy: Italian Society of Pediatric Nutrition/Italian Society of Pediatric Allergy and Immunology Task Force Position Statement. *Ital. J. Pediatr.* **2014**, *40*, 1. [CrossRef] [PubMed]

beverages

MDPI

Review

Raw and Heat-Treated Milk: From Public Health Risks to Nutritional Quality

Francesca Melini [1,2,*,†], Valentina Melini [2,†], Francesca Luziatelli [1] and Maurizio Ruzzi [1]

[1] Department for Innovation in Biological, Agro-food and Forest systems (DIBAF), University of Tuscia,
    Via San Camillo de Lellis snc, I-01100 Viterbo, Italy; f.luziatelli@unitus.it (F.L.); ruzzi@unitus.it (M.R.)
[2] CREA Research Centre for Food and Nutrition, Via Ardeatina 546, I-00178 Roma, Italy;
    valentina.melini@crea.gov.it
*   Correspondence: francesca.melini@gmail.com; Tel.: +39-347-48-14-311
†   These authors contributed equally to this work.

Academic Editor: Alessandra Durazzo
Received: 4 September 2017; Accepted: 28 October 2017; Published: 7 November 2017

Abstract: Consumers have recently shown a preference for natural food products and ingredients and within that framework, their interest in consuming raw drinking milk has been highlighted, claiming nutritional, organoleptic and health benefits. However, a public debate has simultaneously emerged about the actual risks and benefits of direct human consumption of raw milk. This paper compares the microbiological, nutritional and sensory profile of raw and heat-treated milk, to evaluate the real risks and benefits of its consumption. In detail, it provides an updated overview of the main microbiological risks of raw milk consumption, especially related to the presence of pathogens and the main outputs of risk assessment models are reported. After introducing the key aspects of most commonly used milk heat-treatments, the paper also discusses the effects such technologies have on the microbiological, nutritional and sensory profile of milk. An insight into the scientific evidence behind the claimed protective effects of raw milk consumption in lactose-intolerant subjects and against the onset of asthma and allergy disorders in children is provided. The emergence of novel milk processing technologies, such as ohmic heating, microwave heating, high pressure processing, pulsed electric fields, ultrasound and microfiltration is also presented as an alternative to common thermal treatments.

Keywords: raw drinking milk; heat-treated milk; microbiological hazards; risk assessment; lactose intolerance; allergies; milk nutritional quality

1. Introduction

In recent decades, a strong desire for things that are natural has appeared and consumers have shown a preference for natural food products and ingredients. Food naturalness is an abstract construct, hard to define and measure but consumers have interpreted it as synonymous with going shopping at farmers' markets, purchasing organic food, consuming seasonal and minimally processed food products [1]. The demand for fresh-like, nutritious products with high organoleptic quality and an extended shelf-life has hence increased more and more.

A few surveys on the importance of food naturalness for consumers—namely the Kampffmeyer Food Innovation Study in 2012 [2] and the Nielsen Global Health and Wellness Survey in 2015 [3]—have recently been undertaken to understand why freshness, naturalness and minimal processing are the most desirable attributes for a food and what is the drive for that trend. It emerged that natural foods are considered healthier than commercial or processed foods [1].

In this context, the consumption of raw milk and products made from it has appeared. A prevalent belief that milk possesses particularly healthy properties and attributes when it is consumed in its raw form has arisen and, as a result of some perceived health benefits, it has become especially consumed by individuals that may have lowered immunity, such as very young, very old or immune-compromised people, as well as persons with specific dietary habits [4].

Over recent decades, a public debate has, nevertheless, emerged about the actual risks and benefits that direct human consumption of raw milk, as a drinking milk, may have. From a science perspective, food naturalness does not straightforwardly imply food healthiness, tastiness and safety. In fact, 27 milk-borne disease outbreaks occurred from 2007 to 2012 in the EU and an association with the consumption of raw drinking milk was claimed [4].

Recently, the European Food Safety Authority (EFSA) has been called upon to provide scientific opinion on the public health risks related to the consumption of raw drinking milk [4]. The hazards related to raw drinking milk are also well evidenced on the websites of authoritative institutions like the Food and Drug Administration (FDA) [5] and the Centers for Disease Control and Prevention [6].

The aim of this work is to compare the microbiological, nutritional and sensory profile of cow's raw and heat-treated milk in order to evaluate the risks and benefits of its consumption and to provide a basis for driving further research on technologies enabling the production of high quality milk. In detail, an updated overview of the main microbiological risks of raw milk consumption is presented, with emphasis on models for quantitative microbial risk assessment. The most common heat-treatments and their effects on the microbiological, nutritional and sensory profile of milk are reported. The effect of cow's raw milk consumption against lactose intolerance and allergy disease risk is discussed. An overview of novel technologies for milk processing, as an alternative to common heating treatments, is finally presented.

2. Method

2.1. Literature Search

The study layout was first designed and an extensive literature search for papers on planned topics was conducted May to August 2017 by two authors (FM and VM). A limited updated literature search was performed also in October 2017.

Major literature databases were used (i.e., SCOPUS, PubMed, ScienceDirect) in order to identify the literature relevant to the topic. The websites of authoritative Institutions were also consulted: the European Food Safety Authority (https://www.efsa.europa.eu), the U.S. Food and Drug Administration (https://www.fda.gov/), the Center for Disease Control and Prevention (https://www.cdc.gov/), Codex Alimentarius (http://codexalimentarius.org) and the Rapid Alert System for Food and Feed—RASFF (https://ec.europa.eu/food/safety/rasff_en). The legal framework for raw milk sale was searched on the Official Journal of the European Union (EUR-Lex, http://eur-lex.europa.eu/homepage.html).

During the search of the major literature databases, time limits were first set—the year of publication was to be later than 2007 in order to collect the most up-to-date published works. Several combinations of terms were used, depending on the following raw milk-related aspects/topics: consumer behavior/perception; raw milk microbial ecology; raw milk pathogenic microorganisms; risk assessment on raw milk consumption; heat treatment technologies; influence of heat treatments on milk microbiological, nutritional and organoleptic profile; raw milk and lactose intolerance; farm milk and asthma/allergy; and novel milk processing technologies.

2.2. Including and Excluding Criteria

Exclusion was applied to duplicate papers, articles not accessible for authors, or research studies dealing with raw milk other than cow's and with raw milk by-products. Reference lists of articles were also scanned to further identify relevant papers that were not found in electronic databases. The screening of the titles and abstracts was performed by two authors (FM and VM). A screening of the full text resulted in a further exclusion of papers. The key information from the selected papers was reviewed, extracted and grouped in order to meet the scientific requirements of each topic section.

3. Results and Discussion

3.1. Raw Drinking Milk Definition and Legal Framework

According to EU legislation, "raw milk" is milk produced by the secretion of the mammary gland of farmed animals, which has not been heated to more than 40 °C or has not undergone any treatment with an equivalent effect [7]. Raw milk intended for human consumption must be free of pathogens in accordance with the food safety requirements of the General Food Law, i.e., Regulation (EC) No. 178/2002 [8].

Moreover, specific microbial criteria are also laid down by Reg. (EC) 853/2004: for raw cows' milk, ≤100,000 CFU/mL for plate count at 30 °C and ≤400,000 CFU/mL for somatic cells are established, whereas for raw milk from species other than cows, a plate count at 30 °C of ≤1,500,000 CFU/mL is allowed. Health requirements for production animals and hygienic requirements on milk production holdings (e.g., premises and equipment, hygiene conditions during milking, milk collection and transport, staff hygiene) are also regulated, in order to guarantee all the above microbiological characteristics.

Handling and sale of raw milk are also regulated by EU legislation and accurate provisions are given by Regulations (EC) No. 853/2004 and 854/2004 [7,9] within the EU Hygiene package. Only authorized producers registered for supplying raw milk through vending machines are allowed to place raw milk dispensers near the farm or at any other location. Milk has to be cooled to 6 °C at the farm immediately after milking and transferred into dedicated vending machines. Temperature must be held at 0–4 °C between the tank and the nozzle of the vending machine. It is not allowed to stock any milk batch in the vending machine more than 24 h, as certain pathogens are capable of multiplying at low temperatures and prolonged storage may boost their growth. Any residual milk must be removed carefully and the machine must be cleaned prior to refilling. Internal and external cleaning procedures for milk vending machines should be part of good hygienic practices (GHPs).

A key issue often associated with the quality of milk dispensed by vending machines is the formation of bacterial biofilms, which may lead to an increased opportunity for microbial contamination [10–12]. *Listeria monocytogenes* has, for instance, the potential to form biofilms on materials such as stainless steel, rubber, or plastic, which are frequently used in milk handling equipment or tanks [10]. Psychrotrophic strains of *Pseudomonas* spp., *Escherichia coli*, as well as *Bacillus* spores are also involved in formation of biofilms on milk handling equipment. The safety issue is that biofilms may survive the cleaning process. To that aim, special cleaning treatments, e.g., pre-rinsing with water, circulation of sanitizing and/or alkali/acidic solutions and final cleaning with water, are required [13].

Finally, consumers have to take some precautions because raw milk transportation from selling points to home, handling and storage practices can also be a critical point and potentially have an impact on the microbiological quality of raw drinking milk. Insulated bags have to be used to transport raw milk, the duration of transportation to consumption spots must be very short, raw milk is to be boiled before consumption and stored between 0 °C and 4 °C, as indicated on vending machines.

3.2. Microbial Ecology in Raw Milk

Milk is estimated to be sterile in healthy udder cells and not to contain bacteria in the mammary gland at the site of its production, unless there is an intra-mammary infection and/or the animal has a systemic disease. The indigenous flora is mainly represented by the genera *Streptococcus*, *Staphylococcus* and *Micrococcus* (>50% of the overall raw milk flora) [14].

However, as soon as milk is excreted it is immediately colonized by a complex microbiota, which is comprised of a significant population of microorganisms that naturally dwell in the teat skin and the epithelial lining of the teat canal. Specifically, the bovine teat surface is colonized by bacteria belonging to the phylum *Firmicutes* (76%), *Actinobacteria* (4.9%), *Proteobacteria* (17.8%) and *Bacteroides* (1.3%) but also *Planctomycetes*, *Verrucomicrobia*, *Cyanobacteria* and *Chloroflexi* at a lower level [15]. Milking equipment [16], location of the animals [17,18], feeding site [19,20], bedding material [21] and lactation stage [22] have an influence on the raw milk microbiota as well.

The biodiversity of raw milk microbiota is represented by a great variety of species belonging to the domains of bacteria and fungi, and is influenced by the profile of initial microflora but also by the raw milk biochemical composition, the near neutral pH (6.4–6.8) and the high water-activity (a_w), which may contribute to favoring their growth.

Raw milk microbiota can be mainly classified into two main groups: spoilage microorganisms (Table 1) and pathogens (Table 2), which are both undesirable in raw milk. Spoilage microorganisms can, in fact, grow rapidly in milk and alter traits like nutritional and sensory quality. Pathogens present in raw milk are a threat for milk safety and key causative agents of human infections; hence, their presence is very unwanted.

3.2.1. Spoilage Microorganisms

The category of spoilage microorganisms comprises different groups, the main being lactic acid bacteria (LAB), psychrotrophic bacterial populations both Gram-negative (−) and Gram-positive (+), which can grow during milk storage at ≤6 °C, coliforms and fungal populations (both yeasts and molds) [14,16].

Lactic Acid Bacteria

Lactic Acid Bacteria are an integral part of raw milk microbiota [16]. Their biodiversity in milk depends on the kind of milk and other external parameters during milking [14]. In raw ewes' milk, LAB flora is dominated by enterococci (~40%), lactococci (14–20%), leuconostocs (8–18%) and lactobacilli (10–30%); in raw goat's milk it is dominated by lactobacilli [14]. However, lactococci and lactobacilli are usually the most frequently identified LAB, thereof *Lactococcus lactis*, *Lactobacillus brevis* and *Lactobacillus fermentum* are the most frequently found species [23]. *Lactobacillus* spp. also have proteolytic activity and can produce aroma compounds and exopolysaccharides.

LAB role as spoilage microorganisms results from their being acid-producing fermentative agents when storage temperatures are sufficiently high for them to outgrow psychrotrophs or Gram-negative aerobic organisms are inhibited [24].

Psychrotrophic Microorganisms

Milk freshly drawn from the udder often does not contain detectable populations of psychrotrophic bacteria [25]. However, after milk collection psychrotrophic bacteria grow also when the cold chain is applied. Despite these microorganisms have optimal and maximal growth temperatures above 15 and 20 °C, respectively [25], they have, in fact, the ability to grow at low temperatures, such as 2–7 °C. This means that over time psychrotrophic populations can develop in cold stored raw milk and their presence in the raw milk microbiota can become a matter of concern. The drawback of psychrotroph presence in milk is their ability to produce extracellular enzymes, mainly proteases and lipases, which are responsible for spoiling milk but also dairy products, as the extracellular enzymes can resist pasteurization and even ultra-high temperature processing [26].

The prompt application of a cooling treatment after milking and of cold temperatures for storage, which are a routine practice to control the microbiological quality and safety of raw milk, are thus not effective to have a fall-off in the growth rate of psychrotrophic bacteria.

The number of psychrotrophic bacteria, which develop after milk collection, depends on the storage temperature, time and hygienic conditions. For instance, under unsanitary conditions more than 75% of the total microflora is represented by psychrotrophs, whereas in case of sanitary conditions the number of psychrotrophic microorganisms is lower than 10% [26].

Psychrotrophic bacteria from numerous genera have been isolated from raw milk. They are represented primarily by the Gram-negative genera *Pseudomonas*, *Aeromonas*, *Serratia*, *Acinetobacter*, *Alcaligenes*, *Achromobacter*, *Enterobacter*, *Chryseobacterium* and *Flavobacterium* (Table 1) [16,25–27].

Pseudomonas spp. and *Enterobacter* spp. are the most abundant in cold stored raw milk. It is estimated that Gram-negative microflora accounts for more than 90% of the total psychrotrophic raw milk microflora [14].

The Gram-positive genera *Bacillus, Clostridium, Corynebacterium, Microbacterium, Micrococcus, Streptococcus, Staphylococcus* and *Lactobacillus* are also commonly found in raw milk but they only account for a small proportion of the psychrotrophic microflora [27]. *Bacillus* spp. are also the main predominant spore-forming bacteria, therefore *B. licheniformis, B. cereus, B. subtilis* and *B. megaterium* are the most isolated. *B. cereus* is the most common contaminant [24], but *B. subtilis* and *B. licheniformis* are more heat-resistant than *B. cereus* and they spoil sterilized and UHT milk [14]. In 2006, Heyndrickx's group [28] isolated a very heat-resistant mesophilic species of *Bacillus*—i.e., *Bacillus sporothermodurans*—from UHT milk.

The Gram-positive *Arthrobacter* is claimed to enter from the dairy facility, whereas *Corynebacterium* spp. are reportedly found on the teat surface and in the farm environment [16].

The psychrotrophic microflora of raw milk also comprises pathogens like the Gram-negative *Aeromonas hydrophila* and *Yersinia enterocolitica*—the Gram-positive *L. monocytogenes* and toxin-producing strains of *Bacillus cereus*—whose spores can even survive heat treatments in the range of 65–75 °C.

Coliforms

Coliforms are normally found in raw milk with varying levels [14,29]. Their presence is due to different sources, such as water, plant materials, equipment, dirt and feces. High levels of coliforms (e.g., >1000 CFU/mL) generally indicate unsanitary practices on the farm or inadequate refrigeration but also non-adequate management practices, such as milking machine wash failures and fall-offs in the rate of milking units [29].

Attempts have been made to find a correlation between levels of coliform bacteria and the possibility of public health hazards from raw milk consumption. However, so far, no correlation has been identified. A recent survey carried out in the United States [30] demonstrated that coliform counts are not an index of the presence of *B. cereus, E. coli* O157:H7, *L. monocytogenes* and *Salmonella* spp. and that subsequently coliform testing of raw milk intended for human consumption cannot be used as a reliable tool for public health risk screening [30,31]. Further investigations are thus required.

Fungi

Yeasts and molds can also be an important population in raw milk. They usually originate from contaminated environment of the dairy farm and/or processing plant but can also derive from the physiological state of the animal, feeding and climatic conditions [16]. The most commonly detected yeasts in raw milk belong to the genera *Candida, Cryptococcus, Debaryomyces, Geotrichum, Kluyveromyces, Pichia, Rhodotorula* and *Trichosporon*. *Debaryomyces hansenii, Kluyveromyces marxianus* var. *marxianus* and *Kluyveromyces marxianus* var. *lactis* are of particular interest.

The levels to which molds are present in raw milk are lower than yeasts. The most detected mold genera are *Penicillium, Geotrichum, Aspergillus, Mucor, Rhizomucor, Rhizopus* and *Fusarium* [14,15]. Interestingly, over the last ten years notifications and alerts from the Rapid Alert System for Food and Feed (RASFF) have been issued for raw milk contaminated by mycotoxins from Italy, Hungary and Slovenia [32].

Table 1. Spoilage microorganisms in raw milk [1].

Lactic Acid Bacteria						Psychrotrophs		Fungi	
Lactococcus spp.	Streptococcus spp.	Lactobacillus spp.	Leuconostoc spp.	Propionibacterium spp.	Enterococcus spp.	Gram Positive	Gram Negative	Yeasts	Molds
L. lactis spp. cremoris	S. agalactiae	L. acidophilus	L. mesenteroides	P. acidipropionici	E. durans	Arthrobacter spp.	Achromobacter spp.	Candida spp. C. sake C. parapsilosis C. inconspicua	Aspergillus spp.
L. lactis spp. lactis	S. bovis	L. brevis	L. pseudomesenteroides	P. freudenreichii	E. faecalis	Bacillus spp.	Acinetobacter spp.	Cryptococcus spp. C. curvatus C. carnescens C. victoriae	Fusarium spp.
L. piscium	S. gasgalactiae	L. buchneri		P. jensenii	E. faecium	Bifidobacterium	Aeromonas spp.	Debaryomyces hansenii	Geotrichum spp.
L. raffinolactis	S. macedonicus	L. casei		P. thoenii	E. italicus	Brevibacterium	Alcaligenes spp.	Geotrichum spp. G. candidum G. catenulate	Mucor spp.
	S. thermophilus	L. crispatus			E. mundtii	Chlostridium spp.	Chryseobacterium	Kluyveromyces spp. K. marxianus K. lactis	Penicillium spp.
	S. uberis	L. curvatus				Corynebacterium spp.	Enterobacter spp.	Pichia	Rhizomucor
		L. fermentum				Microbacterium	Flavobacterium	Rhodotorula mucilaginosa	Rhizopus
		L. gasseri				Micrococcus	Pseudomonas spp.	Torrubiella	
		L. johnsonii					Serratia	Trichosporon spp. T. cutaneum T. lactis	
		L. paracasei							
		L. pentosus							
		L. plantarum							
		L. reuteri							
		L. rhamnosus							
		L. sake							

[1] The table was prepared on the basis of the information gathered in [14,16,27].

3.2.2. Raw Milk Pathogenic Microorganisms

Raw milk can harbor also a great number of pathogens (Table 2), even when it is sourced from clinically healthy animals and they can represent a serious health threat for humans. They can originate from feed and drinking water (*Toxoplasma gondii*) [33], dairy farm environment (*Salmonella* spp., *L. monocytogenes*, Shiga toxin-producing *E. coli*, *Campylobacter jejuni*, *Y. enterocolitica* and *Clostridium* spp.), mammary gland, cow diseases or infections (*Staphilococcus aureus* and *Brucella* spp.) but also from equipment, raw milk tanks and personnel. Transmission to raw milk can, in fact, occur either from animals (zoonotic pathogens), or from contaminated environment (exogenous pathogens).

Salmonella spp., *Listeria* spp., *E. coli*, *Campylobacter* spp., *Brucella* spp., *Clostridium* spp. and/or *Shigella* spp. are the most common milk-borne pathogens and also the main causative agents of microbial food-borne diseases, specifically, milk-borne infections, milk-borne intoxications and milk-borne toxico-infections [14].

Generally speaking, typical symptoms from drinking raw milk contaminated by the above-mentioned pathogens are fever, nausea, vomiting, diarrhea and abdominal pains. However, they can potentially affect also the cardiovascular, cutaneous, neurological, ocular and pulmonary system, and only in some cases they cause death, as it is the case for *Listeria* spp. (30–35%) and *Streptococcus* spp. (up to 29%) [34].

Salmonella spp. are natural inhabitants of the gastrointestinal trait of animals. Milk contamination by them can generally occur at harvest and only in rare cases they determine sub-clinical mastitis which will cause the milk-borne disease at its turn. They are mesophilic microorganisms with an optimum growth temperature of 35–37 °C [14] but can also grow at a wider temperature range, i.e., 5–46 °C. According to the USA Centre for Disease Control and Prevention (CDC), 38% of raw milk outbreaks involving children from 2007 to 2012 in USA were determined by *Salmonella* spp. [6]. The gastroenteric form of non-typhoid salmonellosis is frequently related to the consumption of raw milk. *Salmonella* spp. have, however, a poor thermal tolerance and are thus sensitive to pasteurization.

Table 2. Main raw milk pathogens and related zoonoses [1].

Pathogen	Taxonomy	Morphology	Disease	Transmission Route	System Potentially Affected					
					Cardio Vascular	Cutaneous	Gastro Intestinal	Neurological	Ocular	Pulmonary
Brucella spp. *B. abortus* *B. melitensis*	Bacteria	Gram (−) coccobacilli	Brucellosis	Cutaneous Ingestion Inhalation	×	×	×	×	×	×
Campylobacter spp. *C. fetus* *C. jejuni*	Bacteria	Gram (−) corkscrew	Campylobacteriosis	Ingestion	×		×	×		
C. burnetii	Bacteria	Gram (−) coccobacilli	Q fever	Ingestion Inhalation	×		×	×		×
E. coli	Bacteria	Gram (−) bacilli	Hemolytic uremic syndrome Hemorrhagiccolitis	Ingestion Inhalation		×	×	×		
L. monocytogenes	Bacteria	Gram (+) bacilli	Listeriosis	Ingestion Cutaneous	×	×	×	×		×
Mycobacterium spp. *M. tuberculosis* *M. bovis*	Bacteria	No Gram classification bacilli	Tuberculosis	Cutaneous Inhalation Ingestion		×	×	×		×
Salmonella spp.	Bacteria	Gram (−) bacilli	Salmonellosis	Ingestion			×			
Shigella spp.	Bacteria	Gram (−) bacilli	Shigellosis	Ingestion		×	×	×		
Staphylococcus spp.	Bacteria	Gram (+) staphylococci	Staphylococcal disease	Cutaneous Inhalation Ingestion	×	×	×	×		×
Streptococcus spp.	Bacteria	Gram (+) streptococci	Toxic shock syndrome	Cutaneous Inhalation Ingestion	×	×	×	×		×
Yersinia spp. *Y. pseudotuberculosis* *Y. enterocolitica*	Bacteria	Gram (−) bacilli	Yersiniosis	Cutaneous Inhalation	×	×	×	×		×

[1] The table was prepared on the basis of the information gathered in [34].

L. monocytogenes is one more example of foodborne pathogen possibly contaminating raw milk. Feces dirtying milking equipment are a source of contamination. In humans, it causes large outbreaks of listeriosis, a serious invasive disease-causing abortion in pregnant women, meningitis, encephalitis, and septicemia in neonates and immune-compromised adults, with quite a general high mortality rate [35]. The threat is due to the fact that it can grow and multiply during raw milk storage also at low temperatures (0–4 °C), that implying that even the application of a correct cold chain would not completely eliminate the microorganism. Its occurrence in raw farm milk and bulk tank milk is reportedly frequent. It can also grow on steel and rubber surfaces and plays an important role in the formation of biofilms in vending machines.

E. coli has been recognized as an indicator of fecal contamination. The most pathogenic strains are referred to as verocyto-toxigenic *E. coli* (VTEC), Shiga toxin-producing *E. coli* (STEC) and enterohemorrhagic *E. coli* (EHEC), also known as *E. coli* serotype O157:H7. Cattle feces are the major reservoir of EHEC which commonly contaminates bulk tank milk. Milk contamination is hence a result of direct exposure to fecal material or environmental contamination. Raw milk poses a risk for STEC, and a number of outbreaks has been recently reported for this pathogen [33,36]. In 2013, 3% of 860 tested raw milk samples were found positive for STEC in Europe [33], whereas in USA, according to CDC, Shiga toxin-producing *E. coli* caused 17% of the outbreaks that occurred 2007 to 2012. VTEC serotypes have also been detected in cow's mastitic milk, that implying that an additional contamination route may be sub-clinical mammary infections. Most strains are not heat-resistant, pasteurization thus destroys them.

Campylobacter spp. belong to the family of *Campylobacteraceae* and are an etiological agent of human gastroenteritis. Among them, the most detected isolate in raw milk is *C. jejuni* which is acid- and heat-sensitive and is hence killed by pasteurization. Outbreaks of campylobacteriosis, following the consumption of raw milk, have been reported in the USA, the Netherlands and Hungary, for instance [14]. Specifically, van Asselt and colleagues (2017) report that raw milk consumption accounts for a relatively high number of *Campylobacter* outbreaks: in 2013, 32 strong-evidence *Campylobacter* spp. outbreaks were reported in the EU, of which between 9% in 2013 and up to 20% in 2012 could be attributed to it [33].

Brucella spp. are the main causative agents of the bacterial zoonosis brucellosis. They are very infectious microorganisms which can cause disease in both animals and humans. The most pathogenic strains which have been associated with disease in humans are *Brucella abortus* and *Brucella melitensis*. The former is more often associated with cattle, whereas *B. melitensis* is especially associated with sheep and goats. Most cases of food-borne brucellosis in humans are contracted via consumption of raw milk and derivatives. Among the milk-borne pathogens, *Brucella* spp. are in fact able to survive and multiply at refrigeration temperatures, together with *L. monocytogenes* and *Y. enterocolitica*. *Brucella* spp. are not particularly resistant to thermal processing and standard pasteurization can sufficiently destroy them. However, the issue with it is that it can survive and multiply in milk also upon contamination after pasteurization.

S. aureus is a Gram-positive bacterium, which causes mastitis in cows and other domestic dairy ruminants. It can come to contaminate milk via the teat canal, when there is infection of the mammary gland, via the environment, or by bad hygiene habits during or after milking, such as, not washing hands when handling milk storage equipment [16]. *S. aureus* may cause diseases through the production of heat-stable enterotoxins. The latter, in fact, are very resistant to heating and pasteurization. For that reason, boiling milk for 1 h may decrease the quantity of toxin present in milk, but autoclaving at 15 psi for 20 min seems to be the main treatment able to completely destroy the toxins [37].

Two more zoonotic bacteria of concern are *Mycobacterium avium* subsp. *Paratubercolosis* (MAP) and *Mycobacterium bovis*. MAP causes para-tuberculosis or Johne's disease, which mainly infect domestic animals. It survives and multiplies in the animal intestinal tract mucosa. In recent times, some evidence has been provided about a relationship between MAP and Crohn's disease in humans [15], though the association remains controversial. High prevalence of MAP has been reported in raw milk. It is however relatively heat-resistant. Dairy processors affirm that it may survive pasteurization at 72 °C for 15 s and trials on its resistance to heat have so far reported controversial results [24]. In 2002, researchers of the Queen's University of Belfast [38] screened 567 samples of commercial pasteurized milk and found that 1.8% were contaminated with *M. avium* subsp. *tuberculosis*. This microorganism can survive HTST pasteurization and can also be found in pasteurized milk due to post-processing contamination.

M. bovis causes bovine tuberculosis in animals but it can also spread to humans via the consumption of raw milk. It causes zoonotic tuberculosis, which is indistinguishable from human tuberculosis. Upon consumption of infected milk, extra pulmonary lesions may develop [36]. However, pasteurization removes it. Moreover, countries like the Netherlands have an official bovine tuberculosis-free status [36] which potentially implies the elimination of the pathogen from the food chain.

Y. enterocolitica is the causative agent of acute gastroenteritis, whose symptoms are abdominal pain, diarrhea and fever. It may however mimic appendicitis and occasionally leads to misdiagnosis. Pasteurization can kill this bacterium; however, it happens sometimes that the heat-treatment is not strong enough or recontamination may occur; the bacterium can thus multiply also under refrigeration temperatures [16]. However, *Y. enterocolitica* incidence in raw milk and low-heat-treated milk products is reportedly low and only a few positive results have been recently reported within the EU [16].

Coxiella burnetii is the causative agent of Q fever. It can infect several animal species, such as cows, sheep, goats, but it is by far the main infectious agent of humans. In them, *C. burnetii* shows up with flu-like symptoms leading to endocarditis and hepatitis. It is relatively heat-resistant but is killed by regular pasteurization treatments.

Ensuring the safety of raw drinking milk can hence be very difficult. The control of handling and storage temperatures can be an approach to maintain the microbiological stability and shelf-life of milk, because for some bacteria present in raw milk higher temperatures are required to grow; however, also when milk is cooled and stored properly at <4 °C, bacterial multiplication is not limited for all bacteria. The growth limitation, for example, is not applicable to psychrotrophic bacterial pathogens which may multiply at these temperatures.

3.3. Assessment of Public Health Risks upon Raw Milk Consumption

3.3.1. Quantitative Microbial Risk Assessment

In situations like this, when public health risks related to consumption of raw milk are claimed by the scientific community and, on the other hand, raw milk is consumed because perceived as more natural than heat-treated milk, the establishment of approaches based on risk analysis is to recommend.

Generally speaking, risk analysis is a process comprising three components: risk assessment, risk management and risk communication [39]. Within a framework of microbiological hazards in foods, risk assessment is the tool by which the quantitative probability of illness cases in population can be expressed.

Quantitative microbial risk assessment (QMRA) provides, in fact, the magnitude of the prevailing risk on the basis of outputs (e.g., epidemiological data) and is the means which enables real public health risks to be assessed by health authorities, so that risk intervention scenarios can be designed and risk management options for implementation of intervention actions can be eventually identified and selected.

Over the last ten years, QMRA models have been developed in Australia [40] and New Zealand [41], in the United States [42,43] and in Europe, specifically in Greece [44], Italy [45–50] and the United Kingdom [51], to evaluate the real risk associated with consumption of raw and/or pasteurized milk.

In details, the predicting models developed to estimate the risk of disease after consumption of raw drinking milk regarded the risk of campylobacteriosis [40,41,45,47], listeriosis [41,43,46], hemolytic uremic syndrome (HUS) [41,45,48], salmonellosis [41,46] and staphylococcal disease [42,50] (Table 3).

Table 3. Microbiological hazards upon raw milk consumption analyzed in the currently available QMRA models.

Reference	Country	Scenarios under Consideration	Campylobacter spp.	L. monocytogenes	Salmonella spp.	S. aureus Staphylococcus enterotoxin A	STEC
[40]	Australia	Farm gate consumption Off-farm sale Sale at retail outlets	✓	✓	✓	-	✓
[41]	New Zealand	Farm gate consumption Farm gate sale Off-farm sale Sale at retail outlets	✓	✓	✓	-	✓
[42]	United States	Pathogen growth and staphilococcal enterotoxin A production scenarios Storage conditions (various times and temperatures)	-	-	-	✓	-
[43]	United States	Farm gate consumption Off-farm sale Sale at retail outlets	-	✓	-	-	-
[45]	Italy	Storage scenario (best and worst storage conditions)	✓ (C. jejuni)	-	✓	-	✓
[46]	Italy	Storage scenario (best and worst storage conditions)	-	✓	✓	-	-
[47]	Italy	Storage scenario (best and worst storage conditions) Boiling and not boiling milk	✓ (C. jejuni)	-	-	-	-
[48]	Italy	Storage scenario (best and worst storage conditions) Boiling and not boiling milk	-	-	-	-	✓
[50]	Italy	Pathogenity of multiple strains of a single pathogen Consumer behavior at household level	-	-	-	✓	-

The models developed by Koutsoumanis and colleagues (2010) [44] and Barker and colleagues (2013) [51] investigated the risk of listeriosis and staphylococcal disease, respectively, in pasteurized milk.

Crotta and colleagues (2016a) assume that storage duration and temperature before consumption might have a critical influence in the final outcomes of the predicting model and thus lead to an overestimation of the risk. They recognize the importance of assessing the risk from farm to table, with a focus on consumer behavior at the household level when milk is no longer under the control of professionals and no enforcement by law can be applied. However, they highlight that the model outputs do have a dependence on the likelihood of a raw milk serving actually being consumed for any storage time-temperature combination. They thus drew the conclusion that ignoring spoilage of raw milk in QMRA models and, hence, assuming that milk is always consumed, regardless of any occurring organoleptic modifications during storage, is not realistic and strongly influences the model outputs [49].

3.3.2. Developed Models

So far, the developed models estimate the risk of illness per serving and/or per year upon consumption of raw milk. All of them were elaborated on the basis of data on prevalence of hazards in raw milk, obtained in previous monitoring actions, as well as on the basis of data on pathogen dose per serving size, dose response and consumption habits. They also take into consideration different scenarios. Consumers can, in fact, obtain raw milk from several sources, and the related pathways are key elements to consider when assessing and managing risk. In some countries, raw milk sale is allowed only on the farm premises, where consumers either bring their own containers and have them filled directly from the bulk tank; and/or in other cases consumers can purchase bottled raw milk from on-farm stores and/or they do it from retail stores. The risk can be also assessed depending on the demographics of the consuming population, e.g., in children, in the intermediate-age population, in perinatal or elderly people, because of a supposed higher susceptibility of some consumers.

These models show a likely link between raw milk consumption and public health risks, especially in some cases, e.g., when some bad storage conditions are applied and/or when the "to boil" indication is neglected. However, due to a lack of epidemiological data, the burden of disease cannot be completely and ever assessed. Further shortcomings in the elaborated predictive models have also emerged—data on the proportion of raw milk sold directly to consumers are limited or lacking; data on the prevalence of hazards in raw milk are lacking; servings are sometimes estimated and not measured; data on the incidence of specific pathogens are based on passive surveillance and underestimate their true incidence.

Moreover, the magnitude of estimates differs by several orders among models, hence any comparison among results might be difficult to be performed.

However, the currently available predicting models can allow to identify the main sources of contamination, highlight the critical points along the milk production and supply chain where contamination is likely to occur the most, and last but not least they can enable the identification of data gaps and control options [4]. By far, improving on-farm hygiene, developing educational programs for consumers might contribute to decreasing the number of predicted cases of illness.

3.4. Heat Treatment of Raw Milk

Milk is exceptionally rich in macronutrients, namely amino acids, lipids and sugar and micronutrients, such as vitamins and minerals. Due to this richness in nutritive components, it is a fertile medium for the growth of microorganisms that may cause milk spoilage and also provoke food-borne diseases in humans. Moreover, enzymes also occur in milk that contribute to the onset of undesirable changes during milk storage. Hence, milk commonly undergoes industrial processing in order to make it safe for human consumption and prolong its shelf-life.

Heat treatment is the most common way of preserving milk and make it safe. The main goals of heating are (i) killing pathogenic microorganisms, (ii) inactivating most (>95%) spoilage organisms and (iii) inactivating enzymes, native to milk or excreted by microorganisms, responsible for the reduction of milk keeping quality.

The most common heat treatments widely used in the dairy industry to achieve milk safety and preservation are pasteurization and UHT (ultra-high temperature) sterilization. Thermization and in-bottle sterilization are also performed on raw milk.

Basically, the above-mentioned heating treatments differ in the heat loads, specifically in the temperature and duration of heating. The choice of the heat treatment to be applied mainly depends on a trade-off among milk safety, extent of milk shelf-life and changes in milk quality. The heat load necessary to achieve milk safety depends, at its turn, on the microbiological quality of raw milk and on the growth potential of spore-forming bacteria after heating. Consumer preferences and target market should be also considered in choosing heating treatments.

Definitely, several combinations of treatment temperature and time can be used and different categories of milk are thus obtained (Table 4).

Table 4. Milk heat-treatments and effects on microbiological, organoleptic and nutritional quality.

Heating Treatment	Heating Conditions	Milk Category	Shelf-Life and Storage Conditions	Microbiological Effect	Nutritional Effect	Organoleptic Effect
HTST pasteurization	72 °C for 15 s (commonly 75 °C for 20 s)	Pasteurized milk	Refrigerated conditions (<7 °C) for 3–21 days based on raw milk quality)	Inactivation of pathogens (included *M. tuberculosis*), molds, yeasts and most bacteria (not all vegetative bacteria are killed).	- Little impact on casein structure; minor changes to whey protein structure; - loss of lysine; - no effect on fatty acid profile; - decrease of most vitamin content but little impact on total dietary intakes thereof; - no effect on milk mineral content and bioavailability.	No heating flavors
High-temperature pasteurization	≥85 °C for 20 s (usually 115–120 °C for 2–5 s)	High-pasteurized milk	Refrigerated conditions (<7 °C) for 45–60 days based on raw milk quality)	- Inactivation of pathogens and all vegetative microorganisms; - Bacterial spores are not killed; - Milk enzymes are not fully inactivated.	- Little impact on casein structure; - denaturation of whey protein structure; - loss of lysine; - no effect on fatty acid profile; - decrease of most vitamin content but little impact on total dietary intakes thereof; - no effect on milk mineral content and bioavailability.	Cooked flavor

Beverages **2017**, *3*, 54

Table 4. *Cont.*

Heating Treatment	Heating Conditions	Milk Category	Shelf-Life and Storage Conditions	Microbiological Effect	Nutritional Effect	Organoleptic Effect
UHT treatment	135–150 °C for 1–4 s (commonly >140 °C for 5 s)	UHT milk	Non-refrigerated conditions (<32 °C) for 3–12 months	- All pathogenic and non-pathogenic microorganisms and spores are destroyed; - Milk enzymes are inactivated; - Some bacterial proteinases and lipases are inactivated.	- Denaturation of whey protein structure; - loss of lysine; - no effect on fatty acid profile; - decrease of most vitamin content but little impact on total dietary intakes thereof; - no effect on milk mineral content and bioavailability.	Cooked and ketone flavor, browning
In-bottle sterilization	105–120 °C for 20–40 min (commonly 110 °C for 30 min)	Sterilized milk	Non-refrigerated conditions (<32 °C) for 8–12 months	- All pathogenic and non-pathogenic microorganisms and spores are destroyed; - Milk enzymes are inactivated; - Some bacterial proteinases and lipases are inactivated.	- Denaturation of whey protein structure; - loss of lysine; - no effect on fatty acid profile; - decrease of most vitamin content but little impact on total dietary intakes thereof; - no effect on milk mineral content and bioavailability.	Sterilized-caramelized flavor and browning

3.4.1. Pasteurization

According to the Codex Alimentarius, pasteurization is a "microbiocidal heat treatment aimed at reducing the number of any pathogenic microorganisms in milk and liquid milk products, if present, to a level at which they do not constitute a significant health hazard. Pasteurization conditions are designed to effectively destroy the organism *Mycobacterium tuberculosis* and *C. burnettii*" [52].

On the basis of the temperature and the time applied, pasteurization can be classified as HTST (high-temperature short-time) pasteurization and LTLT (low-temperature long-time) pasteurization. The former is also referred to as "low pasteurization" and the condition commonly used for milk is 72 °C for 15 s, whereas the LTLT pasteurization is performed at 63 °C for 30 min or at 68 °C for 10 min [52]. Moreover, the HTST pasteurization is carried out as a continuous operation consisting on heating milk in a heat exchanger and holding it for the required time necessary to the destruction and/or inhibition of any hazardous microorganisms. The LTLT pasteurization is performed as a batch operation, that is, milk is placed in a container and then heated to a certain temperature for sufficiently long time to eliminate pathogens.

Heating conditions (temperature and time) depend on the raw milk microbiological quality, on milk fat or sugar content and also vary from country to country based on microorganism strain heat resistance. Thus, pasteurization can be performed also at temperatures higher than 85 °C for 30 s ("high pasteurization"). The increase in temperature and/or an extension of holding time is recommended to inactivate heat-resistant strains of *L. monocytogenes*, *E. coli* and *Campylobacter* spp. [53]. A more intensive heating is also applied to eliminate MAP.

However, a severe heating treatment may affect negatively the keeping quality of milk. For instance, the spores of *Bacillus* spp. may germinate and grow because of heat shocking and the keeping quality of pasteurized milk may be thus reduced [54]. The best keeping quality of pasteurized milk is achieved by using temperatures below 77 °C that do not inactivate the lactoperoxidase enzyme (LPO) and do not stimulate the growth of spores.

3.4.2. UHT Sterilization

The UHT treatment is a sterilization process that has been defined by the Codex Alimentarius as "the application of heat to a continuously flowing product using such high temperature conditions for such time that renders the product commercially sterile at the time of processing. When the UHT treatment is combined with aseptic packaging, it results in a commercially sterile product" [52].

The heating is commonly in the range 135–150 °C for 1 s up to 4 s, in order to achieve "commercial sterility", that is, low probability for microorganisms to grow in the product under the normal conditions of storage.

The UHT process can be performed by "direct" or "indirect" heat transfer. In the direct UHT treatment, superheated steam is mixed with milk. In detail, steam may be injected into milk (steam injection) or milk may be sprayed into steam (steam-infusion). In the indirect system, a heat exchanger transfers heat across a partition between milk and steam or hot water.

One drawback of the indirect method is the possibility of plant to fouling. In the lower-temperature section (<100 °C), whey proteins—mainly β-lactoglobulin—denaturate and deposit, while in the higher-temperature section (>100 °C) calcium phosphate deposits due to its reduced solubility. Thus, a decrease in heat transfer and an increase in pressure might be observed.

The shelf-life of UHT milk may be up to 12 months, despite it is usually consumed much earlier.

3.4.3. In-Bottle Sterilization

In-bottle sterilization is commonly performed at 110 °C for 30 min, however temperatures ranging from 105 °C to 120 °C for 20–40 min can be used [55]. All pathogens and non-pathogens microorganism are destroyed, as well spores. A 9-log reduction in the spores of thermophilic bacteria and 12-log reduction of *C. botulinum* are obtained. All milk enzymes are inactivated but not all bacterial lipases and proteinases.

Nevertheless, this processing shows some drawbacks such as slow product heating and cooling and limitation of temperatures, due to the generated internal pressures. It has also detrimental effect on organoleptic and nutritional quality of milk.

3.4.4. Thermization

Thermization is a heat treatment usually performed at 60 to 69 °C for 20 s. The main purpose is to kill bacteria, especially psychrotrophics, thus preventing the production of heat-resistant lipases and proteinases that may impair the milk keeping quality. Thermization thus enables to extend the storage time of raw milk before processing and to enhance the keeping quality of milk [56]. Nevertheless, it does not ensure milk safety, as it cannot completely eliminate pathogens—*L. monocytogenes* can grow in chilled-stored thermized milk [57] and the effect on *M. bovis* and *C. burnetii* is limited [55].

3.5. Heat Treatment and Milk Quality

Milk heat treatment mainly aims at achieving its safety for human consumption by killing pathogens and/or reducing microorganisms which may cause spoilage. However, changes also in organoleptic and nutritional properties of milk occur during heat treatments depending on the heat load. They are discussed in the following sections.

3.5.1. Microbiological Effect

The microbiota of raw and heat-processed milk deeply differs. As mentioned above, pasteurization was, in fact, conceived to destroy vegetative pathogenic microorganisms, that are the main causative agents of milk-borne diseases.

Salmonella spp., *C. jejuni*, *E. coli*, *L. monocytogenes*, *Y. enterocolitica*, *Brucella* spp. do not generally survive pasteurization (Table 5). Spores of pathogens such as *C. botulinum*, *Clostridium perfringens* and *B. cereus* are, however, not eliminated by heat treatment [56,58], although a very low disease incidence is reported. In particular, the spores of *C. perfringens* do not represent a health hazard in pasteurized milk because they are not able to germinate and grow at refrigeration temperatures. On the other hand, the spores of *B. cereus* can grow at low temperatures and cause milk-borne disease outbreaks.

S. aureus does not survive pasteurization but it may produce heat-stable enterotoxins which are very resistant to heating and pasteurization. In particular, the enterotoxin A can remain active upon heat-treatment at 121 °C for 28 min. Recently, Rall and colleagues (2008) [59] screened raw and pasteurized milk samples for *S. aureus* and found it in 70.4% of raw milk samples, in eight samples of pasteurized milk before the expiration date and in 11 samples analyzed on the expiration date.

The effect of pasteurization on MAP is controversial [24]. According to Ryser (2012), it can survive HTST pasteurization (72 °C for 15 s) and can be present as a post-process contaminant [60]. *M. bovis* is, on the other hand, killed by pasteurization.

Despite *C. burnetii* is the most heat-resistant non-sporulating pathogen present in milk, it does not survive regular pasteurization that was designed to achieve at least a 5-log reduction of *C. burnettii* in whole milk [54].

Milk that has undergone a correct pasteurization treatment is, therefore, unlikely to cause disease [61]. However, in case inadequate heat treatments were applied or recontamination events occurred after pasteurization, *Salmonella* spp., *L. monocytogenes*, *C. jejuni*, *Y. enterocolitica*, STEC, *B. cereus*, *Mycobacterium* spp., *S. aureus*, or *C. botulinum* may be present in milk and dairy products [62,63].

As regards spoilage microorganisms, thermolabile psychrotrophs are killed by pasteurization but post-process contamination and/or heat resistance can occur. For instance, during the filling process pasteurized milk may be contaminated by Gram-negative psychrotrophs. The presence and count of psychrotrophs in pasteurized milk depends on the initial count before the heat-treatment. *Pseudomonas* spp. have been long considered heat-sensitive and unable to survive pasteurization;

however, new analytical methods (culture-independent) have revealed that the *Pseudomonas* population is reduced, rather than eliminated, by pasteurization [15]. This implies that damaged but potentially metabolically active cells are present after the heat treatment. They are hence the most dominant microorganisms present in pasteurized milk, together with *Flavobacterium* which are also present but to a lesser extent. *P. fluorescens* is the main causative agent of off-flavors in milk, e.g., stale, cheesy, sour and bitter [24]. *Lactobacillus* and *Lactococcus* are only rarely found in pasteurized milk. Acidification occurs only when milk is left at room temperature. Low pasteurization also ensures killing of all yeasts and molds that can be in raw milk.

The UHT process destroys all vegetative bacteria (both pathogenic and non-pathogenic) and most spore-formers. However, raw milk quality is a key factor affecting the quality of UHT milk. If a relatively high population of sporeforming bacteria is present in raw milk, a low amount thereof may survive the UHT treatment. For instance, the bacterium *B. sporothermodurans* produces highly-resistant spores [28]. They do not cause spoilage, except a slight discoloration of the milk. However, as reported in a bulletin by the International Dairy Federation, it is very tough to remove them from equipment, and contamination thereof has often been the reason of shutting some UHT plants [24].

Raw milk destined for UHT treatment should be stored at less than 5 °C for no more than 48 h after milking. In the case that raw milk is stored at higher temperatures and/or for longer time, psychrotrophic bacteria may grow as well, and produce lactic acid, which causes a reduction of milk pH and a flat sour defect [24]. Enzymes, such as proteases and lipases, may alter the organoleptic properties of milk, in terms of bitter flavor, gelation and rancid flavor. The UHT process does not inactivate some of the enzymes produced by psychrotrophic bacteria, such as *Pseudomonas* spp. [54].

Heat-resistant thermophiles, like *Geobacillus stearothermophilus* and *B. licheniformis*, might be also encountered in UHT milk; nevertheless, they do not grow in milk stored at less than 30 °C. *Bacillus* spp. can be also found in UHT milk, although it is controversial if their presence is due to post-sterilization contamination or heat-resistance. The most detected species are *B. licheniformis*, *Bacillus coagulans*, *Bacillus badius* and *B. cereus*. The latter is, however, unlikely able to survive UHT treatment [24,64], that implying post-sterilization contamination.

Table 5. Survival of microorganisms to heat treatments.

Microorganisms	Survival to Pasteurization	Survival to UHT
S. aureus	√ (enterotoxins)	√ (enterotoxins)
C. jejuni	×	×
Salmonella spp.	×	×
E. coli	×	×
L. monocytogenes	×	×
Y. enterocolitica	×	×
Mycobacterium avium subsp. *paratubercolosis*	√/×	×
M. bovis	×	×
B. cereus	√ (spores)	×
Clostridium spp.	√ (spores)	√ (spores)

√ = survive; × = not survive; √/× = controversial.

3.5.2. Nutritional Effect

Milk contains nearly all the nutrients necessary to sustain life and because of their balance, milk nutritional value is particularly high. The composition of milk varies, depending on the mammal species, the animal status and health and the feed. Heat treatments also influence the nutritional profile of milk.

In the following sections, the effect of milk processing by heating on nutrients is discussed.

Proteins and Enzymes

Milk contains caseins and whey (or serum) proteins. The former represent 80% of milk proteins, and they are precursors of bioactive compounds with antimicrobial activity. They form micelles, containing calcium and phosphorus. Caseins are not heat labile and they do not undergo heat denaturation (in contrast to whey proteins). However, very severe heat treatments may dephosphorylate, hydrolyze or aggregate them. In detail, they can aggregate, and coagulation can occur. Other factors, such as low milk pH and the Ca^{2+} activity, may determine their coagulation [65].

Whey proteins include α-lactalbumin, β-lactoglobulin, serum albumin, immunoglobulins and bioactive peptides, and have important physiological properties. Heat treatments cause their denaturation, as a consequence serine, serine phosphate, glycosylated serine, cysteine and cysteine residues are formed. These compounds may undergo β-elimination and form dehydroalanine, which can react with several amino acids producing proteins that are not hydrolyzed by the intestinal tract. The nutritional value of milk is thus decreased.

Generally speaking, pasteurization little affects casein structure and causes minor changes to the structure of whey proteins [61,66]. However, no significant changes in the nutritional quality of milk protein due to pasteurization were observed in animal and human studies [67,68]. In contrast, Lacroix and colleagues (2008) observed in a human study that UHT treatment modifies the digestive kinetics and hence the metabolism of dietary proteins [68].

As far as amino acids are concerned, the main essential amino acid in milk is lysine. Heating determines lysine losses, ranging between 1% to 4%, while its effect on the other amino acids is negligible [61]. Lysine losses are caused by the extensive Maillard reaction that takes place during heat treatments, especially in in-bottle sterilization. A partial loss of lysine has been observed in UHT milk also during storage. Nevertheless, the loss of this amino acid is not serious, because in milk protein lysine is in excess [65].

Several indigenous enzymes are also present in milk, and the heating treatments it commonly undergoes can denaturate them. As a consequence, the activity of the enzymatic systems is used as an index of the thermal treatments milk undergoes. The activity of alkaline phosphatase is used to monitor the efficacy of pasteurization; therefore, the inactivation of the enzyme ensures that all non sporeforming pathogens have been killed. The activity of lactoperoxidase is used as an indicator for heat treatments more severe than low pasteurization. Gamma-glutamyl-transferase is also used to detect milk treatment above 77 °C.

Lipids

The fat content of marketed milk is standardized by the removal of cream or the addition of whole milk, semi-skimmed milk or skimmed milk.

During heating treatments, physicochemical and chemical changes of milk lipids may occur. The UHT sterilization may increase the amount of free fatty acids. When the indirect system is used, a higher concentration of free-fatty acids is observed compared to the direct method.

At high temperatures, polyunsaturated fatty acids may be converted into conjugated isomers. It has been observed that conjugated linoleic acid has anti-carcinogenic properties [69]. Recently, Pestana and colleagues (2015) investigated the effects of pasteurization and UHT treatments on milk lipids and found no changes in fat level nor in fatty acid profile [70].

Lactose

Lactose is the main milk carbohydrate. It has prebiotic properties and promotes the absorption of calcium and magnesium.

Pasteurization has no effect on lactose, while treatments at higher temperature, such as UHT sterilization, induce the isomerization to lactulose and the formation of acids and Maillard reaction compounds [61,71,72].

Commonly, the lactulose formation from lactose might be observed via the Lobry de Bruyn-Alberda van Ekestein transformation upon heating under slightly alkaline conditions. Since lactulose is not detectable in raw milk, it is used as a heat load indicator and then as an index of the severity of heat treatment that milk underwent.

Heat treatments above 100 °C also determine the degradation of lactose to acids, especially formic acid and lactic acid, and hence an increase in titratable acidity can be observed.

Lactose may also take part to the Maillard reaction that determines the formation of brown products and flavors.

Vitamins

It has been claimed that raw milk has a higher nutritional value than pasteurized milk since it provides a higher number of vitamins. Actually, the heat treatment conditions, in addition to the packaging type and storage conditions, may affect vitamin content in marketed milk.

Recently, Macdonald and colleagues (2011) performed a systematic review to evaluate the impact of pasteurization on vitamins in raw milk [73]. Forty studies assessing the effects of pasteurization on vitamin content were included, and it was found that vitamin B_{12}, vitamin E, vitamin C, folate and riboflavin (B_2) decreased upon pasteurization. In contrast, vitamin A increased and no significant effect of pasteurization on vitamin B_6 levels were found. Despite some vitamins are destroyed by heat treatments (e.g., vitamin C and folate), the contribution of vitamin content to the recommended daily intake (RDI) should be considered in order to compare the nutritional value of raw and heat-treated milk. For example, 20 L of raw milk per day should be consumed to achieve the vitamin C RDI, therefore the degradation thereof due to heat-treatment is not subject of matter. The same applies to vitamin B_{12} and vitamin E, thereof bovine milk is not an important source in occidental diets with a content of 2–5 μg/L and 10–30 μg/dL, respectively [74]. This implies that the effects of pasteurization on the adult daily intake of these vitamins cannot be a concern in diminishing the nutritive value of milk, just because milk is not a primary source thereof.

This explains also the establishment of food fortification programs for vitamins in milk as a public health intervention in Canada and many other countries to correct and/or prevent nutrition problems of public health significance [75,76].

However, vitamin C protects folic acid from oxidation and its breakdown is connected with that of vitamin B_{12}. As far as vitamin B_{12} is concerned, 250 mL of raw milk contribute to more than 80% to the RDI, while in UHT milk the contribution decreases to about 70% [48].

Minerals

Milk is a good source of some minerals, especially calcium and phosphorous, and no significant differences between raw and heat-treated milk in the content thereof have been reported. Moreover, heat-treatments have no effects on the bioavailability of these nutrients [61].

3.5.3. Organoleptic Effect

Milk of good quality has a slightly sweet taste, a very little odor and a smooth and rich feel in the mouth. It is characterized by whiteness and glossiness.

The heat treatments required to achieve milk safety may influence the organoleptic properties of this food, depending on the heating load. Specifically, they affect the flavor and color of milk.

Each heat treatment causes a distinctive flavor profile. Some flavors are induced by heat treatment, others (caused by microorganisms or enzymes) are reduced or annulled.

The typical "cowy" flavor of fresh milk is reduced or masked due to the formation of flavor compounds, such as cooked flavor, UHT ketone flavor and sterilized-caramelized flavors.

Cooked flavor is mainly caused by the Sulphur compounds originating from denaturation of whey protein. As a matter of fact, the denaturation of whey protein exposes sulfhydryl groups that may form sulfhydrylic acid and dimethyl sulphide. The latter are responsible for the cooked flavor of milk undergone to severe heat treatment, such as high pasteurization and UHT treatment [65]. Nevertheless, freshly processed UHT milk has a "cooked" and "cabbage" flavor that however partly disappears during the first week after processing, due to the oxidation of Sulphur compounds. Indirectly, processed milk has a more intense cooked flavor. In addition, stale flavors may develop due to the higher level of dissolved oxygen [54].

Ketone flavor originating in the lipid fraction is also present in UHT milk.

The caramel-like flavor is also called "sterilized flavor", since it is distinctive of sterilized milk. This flavor is caused by the Maillard reaction that also leads to browning.

Moreover, the organoleptic properties of milk are influenced by storage conditions and by the microbial ecology of heat-treated milk. For instance, microbial growth may cause off-flavors. Psychrotrophic bacteria cause putrid flavors, lactic acid bacteria cause sour flavor, *Bacillus circulans* causes a phenolic flavor in in-bottle sterilized milk, *B. cereus* leads to very unclean flavor [65].

Milk enzymes may contribute to the development of milk flavor. The proteolysis of plasmin in UHT milk leads to a bitter flavor and the lipolysis by lipoprotein lipase causes a rancid flavor in low-pasteurized milk.

3.6. Scientific Evidence behind Claimed Health Benefits of Raw Milk Consumption

The consumption of raw milk has been associated to benefits on human health, such as a higher nutritional value and protection against the lactose intolerance, and asthma and allergy diseases. In contrast, the heating treatment is reported to have detrimental effects on these benefits. The nutritional value of raw milk has been compared to the heat-treated milk in the above sections. Following paragraphs report the role of raw milk in the management of lactose intolerance, and asthma and allergy disorders.

3.6.1. Raw Milk and Lactose Intolerance

Lactose is the main carbohydrate in mammal milk and milk products. The inability to digest lactose is referred to as lactose intolerance and it is due to the lack of the enzyme lactase. The main symptoms include flatulence, bloating, diarrhea and abdominal pain.

The incidence of lactose intolerance increases with age and varies by community and ethnic group [77]. It has been estimated that lactose intolerance affects 65% or more of the total human population. In Asia and in North and South America the percentage of adults unable to tolerate lactose in their diet is very high, while in Ireland and Northern European countries lactose intolerance is rare with 74% to more than 90% of population being lactose tolerant [78].

Recently, raw milk consumption has been claimed to reduce lactose intolerance. It has been suggested that raw milk contains natural lactase enzymes that are not found in heated milk, as they are destroyed by heating. However, a lack of scientific evidence supporting this claim exists. Claeys and colleagues (2013) report that both raw and heated milk contain no lactase, and the production thereof by lactic acid bacteria in raw milk is limited, since raw milk must be stored at refrigerated condition due to safety reasons [61]. In contrast, yogurts are tolerated better than milk, as they contain bacteria having lactase enzyme.

A pilot study on adults positive for lactose malabsorption was recently performed by Mummah and colleagues (2014) in order to assess whether raw milk consumption can reduce lactose intolerance symptoms. No significant differences of intolerance symptoms emerged when subjects consumed raw vs pasteurized milk [79]. Additional studies, possibly with larger study groups, are needed in order to support or refuse the claimed protection of raw milk consumption against lactose intolerance.

3.6.2. Raw Milk and Protection against Asthma and Allergies

Beneficial effects of raw milk consumption on human health have been claimed. Among them, an inverse association between raw milk consumption in childhood and the development of asthma, allergies and atopy has been reported [80].

Asthma and allergies have dramatically increased in last decades, especially in Westernized countries [81]. The raise in the incidence and prevalence of atopic disorders that has been observed over the last 30–40 years has occurred within a time span too short to be explained by a genetic shift in the population, thus environmental and/or lifestyle changes might have significantly contributed to this trend. An increase of asthma prevalence by 50% every decade is reported by Braman (2006) [82] and, as a consequence, the morbidity and mortality rates and economic burden associated with asthma management has raised, as well.

The "hygiene hypothesis" formulated by Strachan (1989) [83] reports an inverse relationship between family size and development of atopic disorders and suggests that a lower incidence of infections in early life could boost the rise in allergic diseases.

Within this hypothesis, it has been supposed that a lifestyle enabling exposure during the childhood to microbes, such as farm-living, may have a protective effect against the onset of allergies. Several European cohort studies focused on the association between farm-living and allergy and asthma in children, namely the European Allergy and Endotoxin (ALEX), the Prevention of Allergy Risk Factors of Sensitization in Children Related to Farming and Anthroposophic Lifestyle (PARSIFAL), and the multidisciplinary study to identify the genetic and environmental causes of asthma in the European community (GABRIEL). Evidence of lower incidence and prevalence of asthma, hay fever and atopic sensitization in children exposed to farming lifestyle has been extensively reported [84–92].

In contrast to the above-mentioned cohort studies, other Authors found farming to be not protective against the development of atopic respiratory disorders [93,94].

The term "farming" actually includes several habits, namely exposure to farm animals, to barns and stables, to endotoxins and to the consumption of farm milk (that is unpasteurized milk). The association between each "farm-factor" and allergy disease risk was evaluated, in order to identify the aspects of farming lifestyle that explain the inverse association. As far as farm milk consumption is concerned, Riedler and colleagues (2001) showed, within the ALEX study, that the consumption of raw milk reduced the development of asthma, hay fever and atopic sensitization and the protection was higher in children younger than one year than in those aged 1–5 years [87]. Perkin and Strachan (2006) investigated the association between different farming factors and the prevalence of allergic disorders in children living in English rural farming and non-farming areas. They found that farmers' children had less current asthma symptoms and seasonal allergic rhinitis but non-eczema symptoms. In contrast, the consumption of unpasteurized milk was associated with less eczema symptoms [95]. Hence, the protective effect was associated with the consumption of unpasteurized milk and was independent of farming status.

Ege and colleagues (2007) also found that farm milk consumption was inversely related with asthma prevalence in children, but pig keeping and frequent stay in sheds also acted as protective factors [96]. Waser and colleagues (2007) showed an inverse association between farm milk consumption and childhood asthma, rhinoconjunctivitis and sensitization to pollen, while other farm produced foods were not related to asthma and allergy prevalence [90]. Data were collected within the PARSIFAL study, a cross-sectional multicenter study including almost 15,000 children aged 5–13 years from five European countries and with different lifestyles: some lived in rural areas, others in (sub)urban area, other had an anthroposophical lifestyle, including restrictive use of antibiotics, antipyretics and vaccinations. As previously observed by Riedler and colleagues (2001) [87], they found that the association between farm milk consumption and development of asthma/allergy was most evident in children consuming farm milk since their first year of life. Unfortunately, results from this survey are based on questionnaire data, and no objective confirmation of the raw milk status of the farm milk is available. Some parents explained they boiled milk prior to consumption, others consumed milk as raw milk.

More recently, Loss and colleagues (2015) studied the effect of consuming raw, boiled and industrially processed milk on common infections in the first year of life, in a prospective cohort study including about 1000 children from rural areas of 5 European countries. It emerged an inverse association between the consumption of raw cow's milk and rhinitis, infections of the respiratory tract, otitis and fever. Boiled farm milk showed a similar but milder effect. Heat-processed milk, except UHT, was found to protect against fever [97].

The timing of exposures also appears to deeply affect the possible protection against allergy disorders. In detail, an early-life exposure was found protective. Besides the evidence in the ALEX study [82], Radon and colleagues (2004) also observed a greater protection from allergy risks by exposure to animal buildings during the first year of life or between ages 3 and 5 [98]. Even the prenatal exposure to farming environment was reported to affect the atopic sensitization at birth. Ege and colleagues (2007) evaluated data from PASTURE cohort study and highlighted that maternal lifestyle during pregnancy, included the use of boiled or un-boiled farm milk, affects the production of fetal IgE, determined in cord blood at birth [99]. In contrast, later life exposures of children do not provide any protection or may exacerbate symptoms [89].

Despite the scientific evidence about protection against allergy disorders, the consumption of raw milk still remains to be discouraged.

3.7. Novel Milk Processing Technologies

Thermal processing of milk is the oldest and most common treatment of raw milk before it is deemed fit for human consumption. However, heat treatments may have some drawbacks, such as changes in the organoleptic properties and lower nutritional value thereof. The increasing demand among consumers for fresh-like products, which are more nutritious and of higher organoleptic quality than heat-treated milk, has led to the emergence of alternative thermal and non-thermal milk processing technologies.

Some of the new technologies can meet these demands. Hence, they have captured the attention of the scientific community, governments, as well as food industries endeavoring to stay one step ahead in terms of technology.

These technologies include among others: ohmic and microwave heating, pulsed electric fields, high hydrostatic pressure, microfiltration and ultrasound.

In novel thermal technologies, such as ohmic heating and microwave heating, rise in temperature in the product is mainly responsible for the effect on milk microbial safety, as in conventional methods. In contrast, non-thermal technologies do not involve heat to kill microorganisms. As a consequence, the detrimental effect that conventional thermal treatment has on milk quality is reduced.

3.7.1. Ohmic Heating

The application of ohmic heating (OH) to milk was known since the 19th century [100], but it fell into disuse due to the high cost of electricity, lack of materials suitable to electrode production, and to difficulties to control the process. Recently, improvements were made and it is currently used to blanch, pasteurize and sterilize milk, vegetable products, fruit preparations and meat products [101–104].

When OH technology is applied to milk, heat is generated directly within milk, by using electrodes contacting the food matrix. The latter actually act as an electrical resistor, thus converting the electrical energy into thermal energy.

All food matrices with an electrical conductivity in the range 0.1–$10\,S\,cm^{-1}$ can always be heated by using OH. The food matrix electrical conductivity increases with temperature; hence the effect of the treatment becomes more effective at higher temperatures [105]. Factors such as, food properties (conductivity, viscosity and specific heat capacity), the design of the treatment equipment and the output of the power supply influence the heating rates.

Dispersed systems show differences in conductivity that may cause a non-uniformity of heating with the formation of hot and cold spots, that are local high and low temperature peaks, respectively. As far as milk is concerned, the liquid phase is the most abundant, and it is also the fraction with the highest electrical conductivity. This results in a faster heating and in a heat dissipation towards the particulate fraction, thus compensating possible non-homogeneities during the heating [106]. Therefore, OH promotes fast and more uniform heating in the food matrix.

The rapid heating has a double advantage: the impact on food quality is reduced and the energy necessary for the treatment is lowered [106], thus resulting a more sustainable technology. A uniform heating also prevents the formation of regions at high or low temperatures that represent a critical point for food quality and food safety, respectively. Hot spots are a food quality problem due to their over processing [107], while cold spots are a food safety issue.

The OH has an additional advantage over the conventional heating: it reduces fouling that reduces the heat transfer rates and promotes the formation of biofilm on surfaces, thus compromising the microbial safety of the final product [108].

Despite the above-mentioned advantages over conventional heating, some issues remain on the way, namely the effects of the OH process on the physical and chemical properties of milk, the cost and difficulties in controlling the process parameters, and the effect thereof on fouling. In addition, so far, the impact of OH process on the allergenicity of milk and dairy products has not been investigated yet [108].

3.7.2. Microwave Heating

Microwave heating (MWH) consists on the use of electromagnetic waves of certain frequencies (300 MHz–300 GHz) to generate heat within products [109]. The heating is caused by the ability of materials to absorb microwave energy and then to convert it into heat.

Microwave heating application derives from the establishment of conditions which, on the one hand, provide the desired degree of safety and, on the other hand, guarantee a minimum product quality degradation.

The application of microwave heating to pasteurize milk has been well studied [110–118] and has been a commercial practice for quite a long time. The industrial setting up of microwave heating processes, nevertheless, faces two major issues. There is a non-uniform temperature distribution inside food product, creating temperature gradients within the product and resulting in hot and cold spots within the food matrix, moreover energy costs are high [119].

One of the key issues in assuring milk safety is, however, how effective the treatment is in inactivating microorganisms of public health concern yet preserving the quality of the product. Overall, pasteurization of milk by MWH can increase milk shelf-life over conventional pasteurized milk due to destruction of psychrotrophic bacteria [120].

According to some studies [110,111,121] heating of milk in a microwave oven at a temperature and time used in normal pasteurization are not successful in inactivating pathogens, such as *Salmonella typhimurium*. Outputs from the studies of Stearns and Vasavada (1986) and Galuska and colleagues (1989) showed that MWH causes sub-lethal injuries to milk-borne pathogens, such as, *L. monocytogenes*, *S. aureus* and *E. coli* [112,116]. Variations in the volume of milk treated by MWH can influence the inactivation of *L. monocytogenes* [111].

Insignificant loss of vitamin A, β-carotene, vitamin B_1 or B_2 also occurs [122]. In detail, Sierra and colleagues (1999) compared the heat stability of vitamins B_1 and B_2 in milk treated by continuous microwave heating and conventional system, and observed no significant losses in the vitamins during microwave heating at 90 °C without holding period; for vitamin B_2 a decrease by 3–5% during 30–60 s of holding was found [123].

A loss of approximately 17% for vitamin E and 36% for vitamin C can be observed [122]. Bai and colleagues (2015) have recently shown that milk layer thickness, microwave time and microwave power can be a significant factor affecting vitamin C concentration with milk layer thickness being the most influencing factor [124].

Microwave heating of milk does not affect fat components. As regards protein compounds, Lopez-Fandino and colleagues (1996) investigated the denaturation of β-lactoglobulin and the inactivation of alkaline phosphatase and lactoperoxidase using a modified microwave oven at

2450 MHz [114]. Upon comparison of the main outcomes of this study with results obtained by conventional thermal treatment in a plate-type heat exchanger, it emerged that the degree of inactivation caused by the thermal treatment was similar in the two cases. More recently, Raman (2007) found that denaturation of whey proteins was lower in milk pasteurized by MWH than by a conventional thermal process, whereas the denaturation of β-lactoglobulin was almost similar in both processes [122].

Contrasting opinions are reported on the influence of MWH on milk sensory profile. According to some studies, volatile components in milk conventionally treated and milk treated by MWH in continuous flow differ significantly [122]; Lopez-Fandino and colleagues (1996), on the other hand, maintained that the sensory characteristics of microwave-pasteurized milk were comparable to those achieved by traditional pasteurization after 15-day storage [114].

3.7.3. Pulsed Electric Field

The application of pulsed electric field (PEF) technology in food processing consists on the treatment of a food matrix, placed between two electrodes, with high voltage (5–20 kV) short (1–10 μs) electric pulses. The use of PEF to milk processing has been reported capable of inactivating unwanted pathogenic and spoilage bacteria while keeping sensory and nutritional attributes unaffected [125]. The inactivation of vegetative forms of microorganisms is due to the formation of hydrophilic pores in the cell membrane and to the opening of protein channels causing the loss of cell membrane functionality.

PEF has been found effective in the inactivation of *Pseudomonas* spp. that constitute the predominant microorganisms limiting the raw milk shelf-life under refrigerated conditions and also of *Listeria* spp., *Salmonella* spp., *E. coli*, *B. cereus*, *S. aureus*, *Brucella* spp., *Coxiella* spp. and *Enterococcus* [126] that can occur in raw milk.

This technology was found to be effective alone or in combination with mild heat treatment. Bermúdez-Aguirre and colleagues (2011) found that mesophilic and psychrophilic bacteria in raw skim milk were inactivated by PEF at 20–40 °C and a synergistic effect of the two treatments was suggested [127]. It also emerged that milk fat content possibly protects the mesophilic and psychrophilic bacteria from inactivation during combined PEF-heat treatment. More recently, McAuley and colleagues (2016) compared the impact of PEF at 53 and 63 °C and conventional heating at 63 °C and 72 °C on raw milk microbiological and physicochemical stability [128]. It emerged that PEF processing (22 μs at 30 kV/cm) at 63 °C achieves microbial stability in whole milk similar to thermal pasteurization (72 °C for 15 s). Compared to latter, PEF combined with heat treatment at 63 °C had no adverse effect on milk physicochemical properties. Moreover, raw milk processing by PEF (22 μs at 30 kV cm^{-1}) at 53 °C extends the shelf-life thereof by 3–4 days in refrigerated conditions (4 °C).

As far as the effect of PEF on milk enzymes is concerned, several studies were performed in order to test the enzyme stability in different dairy products. Buckow and colleagues (2012) investigated the effect of combined PEF/thermal treatments on lactoperoxidase (LPO) dissolved in simulated milk ultra filtrate, and found that LPO inactivation was mainly due to thermal effects; nevertheless, 5–12% inactivation may be related to electro-chemical effects [129]. More recently, Sharma and colleagues (2017) assessed the effect of PEF and thermal treatments both on whole bovine milk enzymes, such as alkaline phosphatase, xanthine oxidase, lipase and plasmin, and on microorganisms count over 21 days of storage at 4 °C [130]. It emerged that the effect of PEF on microorganisms and alkaline phosphatase activity immediately after the treatment and after 21-day storage at 4 °C was comparable to the effect of heat treatment. As far as xanthine oxidase and plasmin are concerned, their activities were reduced after PEF treatment, but at the end of storage period they were similar to raw milk. In addition, the lipolytic activity increased over storage. Hence, it emerged that PEF is suitable to process milk intended for cheese making, since enzymes involved in the development of flavor and aroma are retained.

From a nutritional point of view, it was assessed that the application of PEF 400 µs at 18.3–27.1 kV cm^{-1} did not affect the content of fat soluble vitamins (cholecalciferol and tocopherol) and water-soluble vitamins, except ascorbic acid [125].

Actually, the scale-up of PEF treatment was and still is an engineering challenge, since most studies were performed on small volume samples. Moreover, the scale up should ensure electric field uniformity and consider flow behavior, heat conduction and residence times [128].

3.7.4. High-Pressure Processing

High-Pressure Processing (HPP), also known as High Hydrostatic Pressure (HHP) or Ultra High Pressure (UHP), is a non-thermal treatment representing a clear alternative to traditional heat treatments. It involves application of high pressures at room temperature. Its main advantage is the retention of the food original freshness, color, flavor, taste and nutritional quality, and the non-thermal induction of cooked off-flavors, along with the inactivation of microorganisms. HP treatments are usually performed in the range of 100–1000 MPa at room temperature, or higher when spore inactivation is required (up to 60–80 °C) for up to 30 min [131].

Depending on the microbiological quality of milk, the effect of HPP application at 400–600 MPa may be comparable to pasteurization (72.8 °C, 15 s) [132], whereas it is not to sterilization because of the resistance of spores to HPP.

Application of HPP for microbial inactivation has been extensively studied and reviewed [133]. Overall, pressures ranging between 300 and 600 MPa can be effective in inactivating microorganisms, including foodborne pathogens, without damaging the nutritional and sensory characteristics of food.

Most vegetative forms of microorganisms can be destroyed at 600 MPa for 15 min and at 20–30 °C. Bacteria spores are more resistant to HP than vegetative cells and can survive at a pressure of 1000 MPa. *E. coli* and *L. monocytogenes* are reportedly the most pressure-resistant species at room temperature. Gram-positive microorganisms are more pressure resistant than gram-negative microorganisms. Yeasts and molds are the most sensitive to pressure [134]; most of them are inactivated within a few minutes by 300–400 MPa at 25 °C.

Endospores are far more resistant against HP, requiring treatment at pressures exceeding 1000 MPa and temperature higher than 80 °C for full inactivation [132].

Some authors have demonstrated some difficulties for HP to inactivate microorganisms [134], hence possible combinations of HP have been figured out, such as with mild temperatures (30–50 °C) and/or bacteriocins (nisin, pediocin, lacticin) which sometimes improve the inhibition of foodborne bacteria and spores.

As to milk quality, HPP can have a disruptive effect on milk casein micelles and the structure of whey proteins. β-lactoglobulin are easily denatured under pressure treatments of up to 500 MPa at 25 °C. Denaturation of immunoglobulins and α-lactalbumin occurs at much higher pressures and particularly at 50 °C [114].

Enzyme inactivation by HP is more difficult, as they are more resistant [135]. Resistance to pressures lower than 400 MPa at 25 °C is reported for alkaline phosphatase, lactoperoxidase, phosphohexose-isomerase and γ-glutamyltransferase [114,132].

Small molecules, such as, amino acids, vitamins, simple sugars and flavor components, are not affected by HPP and remain unaltered.

3.7.5. Microfiltration

Microfiltration (MF) is a non-thermal treatment method with the specific advantage of being very effective in the removal of bacterial spores in comparison with conventional pasteurization.

A major drawback of MF is fouling at the membrane surface which affects selectivity in an adverse way and requires frequent rinsing and cleaning procedures which can have a detrimental effect on the cost-effectiveness of the technology [136].

MF offers several opportunities to the dairy chain, as it allows milk products to keep organoleptic characteristics which are similar to fresh milk with improved shelf-life. A good number of micro-filtered milk is available on the market and its success is due to a perceived freshness and the abovementioned extended shelf-life.

3.7.6. Ultrasound

Ultrasounds are waves with a frequency higher than 20 kHz, with a distinction between low- and high-intensity ultrasounds, which have a power level of $\leq$0.1 MHz and 10–1000 W cm^{-2}, respectively [137].

In ultrasonic treatments, ultrasound waves travel though a liquid, alternating compression and expansion cycles. During the expansion cycle, high-intensity ultrasound causes the growth of existing bubbles which implode violently when they attain a volume at which they do not absorb more energy (cavitation phenomenon). At the implosion phase, locally very high temperatures (up to 5500 °C) and pressures (50 MPa) are reached inside the bubbles. This has a detrimental effect on microorganisms.

The main applications of ultrasound in milk and dairy products are due to its effect in inactivating bacteria and enzymes, homogenizing milk, extracting enzymes and lactose hydrolysis [134]. However, it has been stated that the energy consumption required in ultrasound application to kill microorganisms is higher than for conventional methods.

Moreover, it has been demonstrated that ultrasound on its own is not very effective for inactivation of microorganisms and enzymes in milk, hence combinations of ultrasound with heat (thermos-sonication) and pressure (mano-sonication) have been developed [137].

Effects on fat, whey proteins, caseins, alkaline phosphatase, lactoperoxidase and γ-glutamyltransferase have been so far evaluated. Ultrasound continuous-flow system has proved to be an adequate method for preservation and homogenization of milk [138].

Due to the slight effect on microorganisms and enzymes, it has been difficult for ultrasonic treatment to become a commercial process; however, it has a good potential as a minimal processing method in combination with other treatments.

Further investigations are, nevertheless, required to improve the processing equipment and to gain more insights into the effect of ultrasound treatments on milk main components.

4. Conclusions

Raw milk consumption represents a realistic threat for human health and a public risk, because it can act as a vector of pathogens and spoilage microorganisms. Milk processing via heat treatments ensures to achieve milk safety, however it does not completely allow the retention of raw milk primary organoleptic and nutritional characteristics. Good agricultural practices (GAP), good hygienic practices (GHPs) and good animal husbandry practices at the farm level enable to obtain a high quality raw milk, which at its turn allows the application of less severe heat-treatments and thus the preservation of the primary quality of raw milk.

Further investigations are required to explain the claimed protective effect that raw milk has on the onset of asthma and allergy disorders in children. Novel and alternative technologies should be optimized for possible production of industrial milk which is safe and perceived as fresh.

Author Contributions: All authors contributed to the conception and design of the study. Francesca Melini (F.M.) collected the information on milk microbial ecology and hazards and on risk assessment models and wrote the relative paragraphs. Valentina Melini (V.M.) collected the information on milk processing and effects thereof on milk quality and on allergies and wrote the relative sections. F.M. and V.M. jointly worked at the paragraphs on novel milk processing technologies. Francesca Luziatelli critically read the paper. Maurizio Ruzzi read and commented on the paper and contributed expert opinions. All authors read and accepted the final manuscript.

Conflicts of Interest: The authors declare no conflict of interest.

References

1. Román, S.; Sánchez-Siles, L.M.; Siegrist, M. The importance of food naturalness for consumers: Results of a systematic review. *Trends Food Sci. Technol.* **2017**, *67*, 44–57. [CrossRef]
2. GoodMills Innovation. Kampffmeyer Food Innovation Study. 2012. Available online: http://goodmillsinnovation.com/sites/kfi.kampffmeyer.faktor3server.de/files/attachments/1_pi_kfi_cleanlabelstudy_english_final.pdf/ (accessed on 3 September 2017).
3. The Nielsen Company. We Are What We Eat. Healthy Eating Trends around the World. 2015. Available online: https://www.nielsen.com/content/dam/nielsenglobal/eu/nielseninsights/pdfs/Nielsen%20Global%20Health%20and%20Wellness%20Report%20-%20January%202015.pdf (accessed on 3 September 2017).
4. EFSA Panel on Biological Hazards (BIOHAZ). Scientific Opinion on the Public Health Risks Related to the Consumption of Raw Drinking Milk: Public Health Risks Related to Raw Drinking Milk. *EFSA J.* **2015**, *13*, 3940. [CrossRef]
5. U.S. Food and Drug Administration. The Dangers of Raw Milk: Unpasteurized Milk Can Pose a Serious Health Risk. Available online: https://www.fda.gov/food/resourcesforyou/consumers/ucm079516.htm (accessed on 3 September 2017).
6. Centers for Disease Control and Prevention. Raw Milk. Available online: https://www.cdc.gov/foodsafety/rawmilk/raw-milk-index.html (accessed on 3 September 2017).
7. European Parliament and the Council of the European Union. Regulation (EC) No 853/2004 of the European Parliament and of the Council of 29 April 2004 laying down specific hygiene rules for on the hygiene of foodstuffs. *Off. J. Eur. Union* **2004**, *L 139*, 55–205.
8. European Parliament and the Council of the European Union. Regulation (EC) No 178/2002 of the European Parliament and of the Council of 28 January 2002 laying down the general principles and requirements of food law, establishing the European Food Safety Authority and laying down procedures in matters of food safety. *Off. J. Eur. Union* **2002**, *L 31*, 1–24.
9. European Parliament and the Council of the European Union. Regulation (EC) No 854/2004 of the European Parliament and of the Council of 29 April 2004 laying down specific rules for the organisation of official controls on products of animal origin intended for human consumption. *Off. J. Eur. Union* **2004**, *L 139*, 206–320.
10. Latorre, A.A.; Van Kessel, J.S.; Karns, J.S.; Zurakowski, M.J.; Pradhan, A.K.; Boor, K.J.; Jayarao, B.M.; Houser, B.A.; Daugherty, C.S.; Schukken, Y.H. Biofilm in milking equipment on a dairy farm as a potential source of bulk tank milk contamination with *Listeria monocytogenes*. *J. Dairy Sci.* **2010**, *93*, 2792–2802. [CrossRef] [PubMed]
11. Giacometti, F.; Serraino, A.; Finazzi, G.; Daminelli, P.; Losio, M.N.; Tamba, M.; Garigliani, A.; Mattioli, R.; Riu, R.; Zanoni, R.G. Field handling conditions of raw milk sold in vending machines: Experimental evaluation of the behaviour of *Listeria monocytogenes*, *Escherichia coli* O157:H7, *Salmonella Typhimurium* and *Campylobacter jejuni*. *Ital. J. Anim. Sci.* **2012**, *11*, e24. [CrossRef]
12. Marchand, S.; De Block, J.; De Jonghe, V.; Coorevits, A.; Heyndrickx, M.; Herman, L. Biofilm Formation in Milk Production and Processing Environments; Influence on Milk Quality and Safety. *Compr. Rev. Food Sci. Food Saf.* **2012**, *11*, 133–147. [CrossRef]
13. Moatsou, G. Sanitary Procedures, Heat Treatments and Packaging. In *Milk and Dairy Products in Human Nutrition*; Park, Y.W., Haenlein, G.F.W., Eds.; John Wiley & Sons: Chichester, UK, 2013; pp. 288–309.
14. Moatsou, G.; Moschopoulou, E. Microbiology of Raw Milk. In *Dairy Microbiology and Biochemistry: Recent Developments*; Ozer, B.H., Akdemir-Evrendilek, G., Eds.; CRC Press—Taylor & Francis Group: Boca Raton, FL, USA, 2015; pp. 1–38.
15. Quigley, L.; McCarthy, R.; O'Sullivan, O.; Beresford, T.P.; Fitzgerald, G.F.; Ross, R.P.; Stanton, C.; Cotter, P.D. The microbial content of raw and pasteurized cow milk as determined by molecular approaches. *J. Dairy Sci.* **2013**, *96*, 4928–4937. [CrossRef] [PubMed]
16. Quigley, L.; O'Sullivan, O.; Stanton, C.; Beresford, T.P.; Ross, R.P.; Fitzgerald, G.F.; Cotter, P.D. The complex microbiota of raw milk. *FEMS Microbiol. Rev.* **2013**, *37*, 664–698. [CrossRef] [PubMed]

17. Bonizzi, I.; Buffoni, J.N.; Feligini, M.; Enne, G. Investigating the relationship between raw milk bacterial composition, as described by intergenic transcribed spacer-PCR fingerprinting and pasture altitude. *J. Appl. Microbiol.* **2009**, *107*, 1319–1329. [CrossRef] [PubMed]
18. Vacheyrou, M.; Normand, A.-C.; Guyot, P.; Cassagne, C.; Piarroux, R.; Bouton, Y. Cultivable microbial communities in raw cow milk and potential transfers from stables of sixteen French farms. *Int. J. Food Microbiol.* **2011**, *146*, 253–262. [CrossRef] [PubMed]
19. Hagi, T.; Kobayashi, M.; Nomura, M. Molecular-based analysis of changes in indigenous milk microflora during the grazing period. *Biosci. Biotechnol. Biochem.* **2010**, *74*, 484–487. [CrossRef] [PubMed]
20. Van Hoorde, K.; Heyndrickx, M.; Vandamme, P.; Huys, G. Influence of pasteurization, brining conditions and production environment on the microbiota of artisan Gouda-type cheeses. *Food Microbiol.* **2010**, *27*, 425–433. [CrossRef] [PubMed]
21. Von Neubeck, M.; Baur, C.; Krewinkel, M.; Stoeckel, M.; Kranz, B.; Stressler, T.; Fischer, L.; Hinrichs, J.; Scherer, S.; Wenning, M. Biodiversity of refrigerated raw milk microbiota and their enzymatic spoilage potential. *Int. J. Food Microbiol.* **2015**, *211*, 57–65. [CrossRef] [PubMed]
22. Callon, C.; Duthoit, F.; Delbès, C.; Ferrand, M.; Le Frileux, Y.; De Crémoux, R.; Montel, M.-C. Stability of microbial communities in goat milk during a lactation year: Molecular approaches. *Syst. Appl. Microbiol.* **2007**, *30*, 547–560. [CrossRef] [PubMed]
23. Bluma, A.; Ciprovica, I. Diversity of lactic acid bacteria in raw milk. In *Research for Rural Development, Proceedings of the International Scientific Conference: Research for Rural Development, Jelgava, Latvia, 13–15 May 2015*; Latvia University of Agriculture: Jelgava, Latvia, 2015.
24. Touch, V.; Deeth, H.C. Microbiology of Raw and Market Milks. In *Milk Processing and Quality Management*; Tamine, A.Y., Ed.; Wiley-Blackwell: Oxford, UK, 2009; pp. 48–71.
25. De Oliveira, G.B.; Favarin, L.; Luchese, R.H.; McIntosh, D. Psychrotrophic bacteria in milk: How much do we really know? *Braz. J. Microbiol.* **2015**, *46*, 313–321. [CrossRef] [PubMed]
26. Hantsis-Zacharov, E.; Halpern, M. Culturable Psychrotrophic Bacterial Communities in Raw Milk and Their Proteolytic and Lipolytic Traits. *Appl. Environ. Microbiol.* **2007**, *73*, 7162–7168. [CrossRef] [PubMed]
27. Vithanage, N.R.; Dissanayake, M.; Bolge, G.; Palombo, E.A.; Yeager, T.R.; Datta, N. Biodiversity of culturable psychrotrophic microbiota in raw milk attributable to refrigeration conditions, seasonality and their spoilage potential. *Int. Dairy J.* **2016**, *57*, 80–90. [CrossRef]
28. Scheldeman, P.; Herman, L.; Foster, S.; Heyndrickx, M. *Bacillus* sporothermodurans and other highly heat-resistant spore formers in milk. *J. Appl. Microbiol.* **2006**, *101*, 542–555. [CrossRef] [PubMed]
29. Martin, N.H.; Trmčić, A.; Hsieh, T.-H.; Boor, K.J.; Wiedmann, M. The Evolving Role of Coliforms as Indicators of Unhygienic Processing Conditions in Dairy Foods. *Front. Microbiol.* **2016**, *7*. [CrossRef] [PubMed]
30. Jackson, E.E.; Erten, E.S.; Maddi, N.; Graham, T.E.; Larkin, J.W.; Blodgett, R.J.; Schlesser, J.E.; Reddy, R.M. Detection and enumeration of four foodborne pathogens in raw commingled silo milk in the United States. *J. Food Prot.* **2012**, *75*, 1382–1393. [CrossRef] [PubMed]
31. D'Amico, D.J.; Groves, E.; Donnelly, C.W. Low incidence of foodborne pathogens of concern in raw milk utilized for farmstead cheese production. *J. Food Prot.* **2008**, *71*, 1580–1589. [CrossRef] [PubMed]
32. Rapid Alert System for Food and Feed. Available online: https://ec.europa.eu/food/safety/rasff_en (accessed on 3 September 2017).
33. Van Asselt, E.D.; van der Fels-Klerx, H.J.; Marvin, H.J.P.; van Bokhorst-van de Veen, H.; Groot, M.N. Overview of Food Safety Hazards in the European Dairy Supply Chain. *Compr. Rev. Food Sci. Food Saf.* **2017**, *16*, 59–75. [CrossRef]
34. McDaniel, C.J.; Cardwell, D.M.; Moeller, R.B.; Gray, G.C. Humans and Cattle: A Review of Bovine Zoonoses. *Vector Borne Zoonotic Dis.* **2014**, *14*, 1–19. [CrossRef] [PubMed]
35. Hunt, K.; Drummond, N.; Murphy, M.; Butler, F.; Buckley, J.; Jordan, K. A case of bovine raw milk contamination with *Listeria monocytogenes*. *Ir. Vet. J.* **2012**, *65*, 13. [CrossRef] [PubMed]
36. O'Mahony, M.; Fanning, S.; Whyte, P. The Safety of Raw Liquid Milk. In *Milk Processing and Quality Management*; Tamine, A.Y., Ed.; Wiley-Blackwell: Oxford, UK, 2009; pp. 139–167.
37. Dhanashekar, R.; Akkinepalli, S.; Nellutla, A. Milk-borne infections. An analysis of their potential effect on the milk industry. *Germs* **2012**, *2*, 101–109. [CrossRef] [PubMed]

38. Grant, I.R.; Ball, H.J.; Rowe, M.T. Incidence of *Mycobacterium paratuberculosis* in bulk raw and commercially pasteurized cows' milk from approved dairy processing establishments in the United Kingdom. *Appl. Environ. Microbiol.* **2002**, *68*, 2428–2435. [CrossRef] [PubMed]

39. World Health Organisation. Food and Agriculture Organisation and Codex Alimentarius Commission. Principles and Guidelines for the Conduct of Microbiological Risk Management (MRM). 2007. Available online: https://www.google.it/url?sa=t&rct=j&q=&esrc=s&source=web&cd=1&cad=rja&uact=8&ved=0ahUKEwjsrbf7lYbXAhVBbhQKHQ0fD7YQFggyMAA&url=http%3A%2F%2Fwww.fao.org%2Finput%2Fdownload%2Fstandards%2F10741%2FCXG_063e.pdf&usg=AOvVaw16HPgG3XDCD5t7PyRqEt9B (accessed on 8 October 2017).

40. Food Standards Australia New Zealand (FSANZ). Microbiological Risk Assessment of Raw Cow Milk. Risk Assessment Microbiology Section. December 2009. Available online: https://www.google.it/url?sa=t&rct=j&q=&esrc=s&source=web&cd=1&ved=0ahUKEwjigfTNnevWAhXKEVAKHc8ECMMQFggnMAA&url=https%3A%2F%2Fwww.foodstandards.gov.au%2Fcode%2Fproposals%2Fdocuments%2Fp1007%2520ppps%2520for%2520raw%2520milk%25201ar%2520sd1%2520cow%2520milk%2520risk%2520assessment.pdf&usg=AOvVaw0XYHQ27rcYxv4ld8jkBqkH (accessed on 12 October 2017).

41. Soboleva, T. Assessment of the Microbiological Risks Associated with the Consumption of Raw Milk. Ministry for Primary Industries (MPI) Technical Paper No: 2014/12. June 2013. Available online: https://www.google.it/url?sa=t&rct=j&q=&esrc=s&source=web&cd=1&cad=rja&uact=8&ved=0ahUKEwjG5sfesevWAhXGaVAKHXycCFkQFggnMAA&url=https%3A%2F%2Fwww.mpi.govt.nz%2Fdmsdocument%2F1118-assessment-of-the-microbiological-risks-associated-with-the-consumption-of-raw-milk&usg=AOvVaw3Wmo0Ycg1gk84D9lavhx6n (accessed on 2 October 2017).

42. Heidinger, J.C.; Winter, C.K.; Cullor, J.S. Quantitative microbial risk assessment for Staphylococcus aureus and Staphylococcus enterotoxin A in raw milk. *J. Food Prot.* **2009**, *72*, 1641–1653. [CrossRef] [PubMed]

43. Latorre, A.A.; Pradhan, A.K.; Van Kessel, J.A.; Karns, J.S.; Boor, K.J.; Rice, D.H.; Mangione, K.J.; Gröhn, Y.T.; Schukken, Y.H. Quantitative risk assessment of listeriosis due to consumption of raw milk. *J. Food Prot.* **2011**, *74*, 1268–1281. [CrossRef] [PubMed]

44. Koutsoumanis, K.; Pavlis, A.; Nychas, G.-J.E.; Xanthiakos, K. Probabilistic Model for *Listeria monocytogenes* Growth during Distribution, Retail Storage and Domestic Storage of Pasteurized Milk. *Appl. Environ. Microbiol.* **2010**, *76*, 2181–2191. [CrossRef] [PubMed]

45. Giacometti, F.; Serraino, A.; Bonilauri, P.; Ostanello, F.; Daminelli, P.; Finazzi, G.; Losio, M.N.; Marchetti, G.; Liuzzo, G.; Zanoni, R.G.; et al. Quantitative risk assessment of verocytotoxin-producing *Escherichia coli* O157 and *Campylobacter jejuni* related to consumption of raw milk in a province in Northern Italy. *J. Food Prot.* **2012**, *75*, 2031–2038. [CrossRef] [PubMed]

46. Giacometti, F.; Bonilauri, P.; Albonetti, S.; Amatiste, S.; Arrigoni, N.; Bianchi, M.; Bertasi, B.; Bilei, S.; Bolzoni, G.; Cascone, G.; et al. Quantitative risk assessment of human salmonellosis and listeriosis related to the consumption of raw milk in Italy. *J. Food Prot.* **2015**, *78*, 13–21. [CrossRef] [PubMed]

47. Giacometti, F.; Bonilauri, P.; Amatiste, S.; Arrigoni, N.; Bianchi, M.; Losio, M.N.; Bilei, S.; Cascone, G.; Comin, D.; Daminelli, P.; et al. Human campylobacteriosis related to the consumption of raw milk sold by vending machines in Italy: Quantitative risk assessment based on official controls over four years. *Prev. Vet. Med.* **2015**, *121*, 151–158. [CrossRef] [PubMed]

48. Giacometti, F.; Bonilauri, P.; Piva, S.; Scavia, G.; Amatiste, S.; Bianchi, D.M.; Losio, M.N.; Bilei, S.; Cascone, G.; Comin, D.; et al. Paediatric HUS Cases Related to the Consumption of Raw Milk Sold by Vending Machines in Italy: Quantitative Risk Assessment Based on *Escherichia coli* O157 Official Controls over 7 years. *Zoonoses Pub. Health* **2016**, *64*, 505–516. [CrossRef] [PubMed]

49. Crotta, M.; Paterlini, F.; Rizzi, R.; Guitian, J. Consumers' behavior in quantitative microbial risk assessment for pathogens in raw milk: Incorporation of the likelihood of consumption as a function of storage time and temperature. *J. Dairy Sci.* **2016**, *99*, 1029–1038. [CrossRef] [PubMed]

50. Crotta, M.; Rizzi, R.; Varisco, G.; Daminelli, P.; Cunico, E.C.; Luini, M.; Grober, H.U.; Paterlini, F.; Guitian, J. Multiple-Strain Approach and Probabilistic Modeling of Consumer Habits in Quantitative Microbial Risk Assessment: A Quantitative Assessment of Exposure to Staphylococcal Enterotoxin A in Raw Milk. *J. Food Prot.* **2016**, *79*, 432–441. [CrossRef] [PubMed]

51. Barker, G.C.; Goméz-Tomé, N. A risk assessment model for enterotoxigenic Staphylococcus aureus in pasteurized milk: A potential route to source-level inference. *Risk Anal. Off. Publ. Soc. Risk Anal.* **2013**, *33*, 249–269. [CrossRef] [PubMed]

52. Codex Alimentarius. Standard CAC-RCP57-2004: Code on Hygienic Practice for Milk and Milk Products. 2004. Available online: http://codexalimentarius.org (accessed on 31 August 2017).

53. Kelly, A.L.; O'Shea, N. Plant and Equipment—Pasteurizers, Design and Operation. In *Encyclopedia of Dairy Sciences*; Academic Press: Cambridge, MA, USA, 2011.

54. Tamime, A.Y. *Milk Processing and Quality Management*; Wiley-Blackwell: Oxford, UK, 2009.

55. Ozer, B.; Akdemir-Evrendilek, G. *Dairy Microbiology and Biochemistry: Recent Developments*; CRC Press—Taylor & Francis Group: Boca Raton, FL, USA, 2015.

56. Kelly, A.; Datta, N.; Deeth, H. Thermal Processing of Dairy Products. In *Thermal Food Processing: New Technologies and Quality Issues. Contemporary Food Engineering*; CRC Press—Taylor & Francis Group: Boca Raton, FL, USA, 2012; pp. 273–306.

57. Fernandes, R. *Microbiology Handbook: Dairy Products*; Leatherhead Pub.: Leatherhead, UK; Royal Society of Chemistry: Cambridge, UK, 2009.

58. Papademas, P.; Bintsis, T. Food Safety Management Systems (FSMS) in the Dairy Industry: A Review. *Int. J. Dairy Technol.* **2010**, *63*, 489–503. [CrossRef]

59. Rall, V.L.M.; Vieira, F.P.; Rall, R.; Vieitis, R.L.; Fernandes, A.; Candeias, J.M.G.; Cardoso, K.F.G.; Araújo, J.P. PCR detection of staphylococcal enterotoxin genes in Staphylococcus aureus strains isolated from raw and pasteurized milk. *Vet. Microbiol.* **2008**, *132*, 408–413. [CrossRef] [PubMed]

60. Ryser, E.T. Safety of Dairy Products. In *Microbial Food Safety*; Food Science Text Series; Springer: New York, NY, USA, 2012; pp. 127–145.

61. Claeys, W.L.; Cardoen, S.; Daube, G.; De Block, J.; Dewettinck, K.; Dierick, K.; De Zutter, L.; Huyghebaert, A.; Imberechts, H.; Thiange, P.; et al. Raw or heated cow milk consumption: Review of risks and benefits. *Food Control* **2013**, *31*, 251–262. [CrossRef]

62. Braunig, J.; Hall, P. Milk and dairy products. In *Micro-Organisms in Foods*; Roberts, T.A., Cordier, J.L., Gram, L., Tompkin, R.B., Pitt, J.I., Gorris, L.G.M., Swanson, K.M.J., Eds.; Kluwer Academic/Plenum Publishers: New York, NY, USA, 2005; pp. 643–715.

63. Farrokh, C.; Jordan, K.; Auvray, F.; Glass, K.; Oppegaard, H.; Raynaud, S.; Thevenot, D.; Condron, R.; De Reu, K.; Govaris, A.; et al. Review of Shiga-toxin-producing *Escherichia coli* (STEC) and their significance in dairy production. *Int. J. Food Microbiol.* **2013**, *162*, 190–212. [CrossRef] [PubMed]

64. Simmonds, P.; Mossel, B.L.; Intaraphan, T.; Deeth, H.C. Heat resistance of *Bacillus* spores when adhered to stainless steel and its relationship to spore hydrophobicity. *J. Food Prot.* **2003**, *66*, 2070–2075. [CrossRef] [PubMed]

65. Walstra, P.; Walstra, P.; Wouters, J.T.M.; Geurts, T.J. *Dairy Science and Technology*, 2nd ed.; CRC Press: Boca Raton, FL, USA, 2005.

66. Braun-Fahrländer, C.; Von Mutius, E. Can farm milk consumption prevent allergic diseases? *Clin. Exp. Allergy* **2011**, *41*, 29–35. [CrossRef] [PubMed]

67. Lacroix, M.; Bos, C.; Léonil, J.; Airinei, G.; Luengo, C.; Daré, S.; Benamouzig, R.; Fouillet, H.; Fauquant, J.; Tomé, D.; et al. Compared with casein or total milk protein, digestion of milk soluble proteins is too rapid to sustain the anabolic postprandial amino acid requirement. *Am. J. Clin. Nutr.* **2006**, *84*, 1070–1079. [PubMed]

68. Lacroix, M.; Bon, C.; Bos, C.; Léonil, J.; Benamouzig, R.; Luengo, C.; Fauquant, J.; Tomé, D.; Gaudichon, C. Ultra high temperature treatment but not pasteurization, affects the postprandial kinetics of milk proteins in humans. *J. Nutr.* **2008**, *138*, 2342–2347. [CrossRef] [PubMed]

69. Fox, P.F.; McSweeney, P.L.H. *Dairy Chemistry and Biochemistry*; Chapman and Hall: London, UK, 1998.

70. Pestana, J.M.; Gennari, A.; Wissmann Monteiro, B.; Neutzling Lehn, D.; Volken de Souza, C.F. Effects of Pasteurization and Ultra-High Temperature Processes on Proximate Composition and Fatty Acid Profile in Bovine Milk. *Am. J. Food Technol.* **2015**, *10*, 265–272. [CrossRef]

71. Ijaz, N. Epidemiological Hazard Characterization and Risk Assessment for Unpasteurized Milk Consumption: United States, 1998–2010; Working Paper; 2013. Available online: https://www.google.it/url?sa=t&rct=j&q=&esrc=s&source=web&cd=1&ved=0ahUKEwjyyaSpyJ_XAhVJ56QKHeH5DJwQFggnMAA&url=http%3A%2F%2Fwww.bccdc.ca%2FHealth-Professionals-Site%2F_layouts%2F15%2FDocIdRedir.aspx%3FID%3DBCCDC-291-107&usg=AOvVaw0cqohwDQiqIKMyh80gd24B (accessed on 6 November 2017).

72. Lejeune, J.; Rajala-Schults, P.J. Unpasteurized milk: A continued public health threat. *Clin. Infect. Dis.* **2009**, *48*, 93–100. [CrossRef] [PubMed]
73. Macdonald, L.E.; Brett, J.; Kelton, D.; Majowicz, S.E.; Snedeker, K.; Sargeant, J.M. A systematic review and meta-analysis of the effects of pasteurization on milk vitamins and evidence for raw milk consumption and other health-related outcomes. *J. Food Prot.* **2011**, *74*, 1814–1832. [CrossRef] [PubMed]
74. Jensen, R.G. *Handbook of Milk Composition*; Academic Press: San Diego, CA, USA, 1995.
75. Canada Department of Justice. *Food and Drug Regulations, Part D, Division 3. Addition of Vitamins, Mineral Nutrients or Amino Acids to Foods*. 2017. Available online: http://laws-lois.justice.gc.ca/eng/regulations/c.r. c.,_c._870/page-144.html (accessed on 22 October 2017).
76. European Parliament; Council of the European Union. Regulation (EC) No 1925/2006 of the European Parliament and of the Council of 20 December 2006 on the addition of vitamins and minerals and of certain other substances to foods. *Off. J. Eur. Union* **2006**, *L 404*, 26–38.
77. Law, D.; Conklin, J.; Pimentel, M. Lactose intolerance and the role of the lactose breath test. *Am. J. Gastroenterol.* **2010**, *105*, 1726–1728. [CrossRef] [PubMed]
78. Vuorisalo, T.; Arjamaa, O.; Vasemägi, A.; Taavitsainen, J.-P.; Tourunen, A.; Saloniemi, I. High lactose tolerance in North Europeans: A result of migration, not in situ milk consumption. *Perspect. Biol. Med.* **2012**, *55*, 163–174. [CrossRef] [PubMed]
79. Mummah, S.; Oelrich, B.; Hope, J.; Vu, Q.; Gardner, C.D. Effect of raw milk on lactose intolerance: A randomized controlled pilot study. *Ann. Fam. Med.* **2014**, *12*, 134–141. [CrossRef] [PubMed]
80. Brick, T.; Schober, Y.; Böcking, C.; Pekkanen, J.; Genuneit, J.; Loss, G.; Dalphin, J.-C.; Riedler, J.; Lauener, R.; Nockher, W.A.; et al. ω-3 fatty acids contribute to the asthma-protective effect of unprocessed cow's milk. *J. Allergy Clin. Immunol.* **2016**, *137*, 1699–1706.e13. [CrossRef] [PubMed]
81. Brooks, C.; Pearce, N.; Douwes, J. The hygiene hypothesis in allergy and asthma: An update. *Curr. Opin. Allergy Clin. Immunol.* **2013**, *13*, 70–77. [CrossRef] [PubMed]
82. Braman, S.S. The global burden of asthma. *Chest* **2006**, *130*, 4S–12S. [CrossRef] [PubMed]
83. Strachan, D.P. Hay fever, hygiene and household size. *BMJ* **1989**, *299*, 1259–1260. [CrossRef] [PubMed]
84. Kilpeläinen, M.; Terho, E.O.; Helenius, H.; Koskenvuo, M. Farm environment in childhood prevents the development of allergies. *Clin. Exp. Allergy J. Br. Soc. Allergy Clin. Immunol.* **2000**, *30*, 201–208. [CrossRef]
85. Riedler, J.; Eder, W.; Oberfeld, G.; Schreuer, M. Austrian children living on a farm have less hay fever, asthma and allergic sensitization. *Clin. Exp. Allergy* **2000**, *30*, 194–200. [CrossRef] [PubMed]
86. Von Ehrenstein, O.S.; Von Mutius, E.; Illi, S.; Baumann, L.; Böhm, O.; von Kries, R. Reduced risk of hay fever and asthma among children of farmers. *Clin. Exp. Allergy* **2000**, *30*, 187–193. [CrossRef] [PubMed]
87. Riedler, J.; Braun-Fahrländer, C.; Eder, W.; Schreuer, M.; Waser, M.; Maisch, S.; Carr, D.; Schierl, R.; Nowak, D.; von Mutius, E.; et al. Exposure to farming in early life and development of asthma and allergy: A cross-sectional survey. *Lancet Lond. Engl.* **2001**, *358*, 1129–1133. [CrossRef]
88. Braun-Fahrländer, C.; Riedler, J.; Herz, U.; Eder, W.; Waser, M.; Grize, L.; Maisch, S.; Carr, D.; Gerlach, F.; Bufe, A.; et al. Environmental exposure to endotoxin and its relation to asthma in school-age children. *N. Engl. J. Med.* **2002**, *347*, 869–877. [CrossRef] [PubMed]
89. Naleway, A.L. Asthma and Atopy in Rural Children: Is Farming Protective? *Clin. Med. Res.* **2004**, *2*, 5–12. [CrossRef] [PubMed]
90. Waser, M.; Michels, K.B.; Bieli, C.; Flöistrup, H.; Pershagen, G.; von Mutius, E.; Ege, M.; Riedler, J.; Schram-Bijkerk, D.; Brunekreef, B.; et al. Inverse association of farm milk consumption with asthma and allergy in rural and suburban populations across Europe. *Clin. Exp. Allergy J. Br. Soc. Allergy Clin. Immunol.* **2007**, *37*, 661–670. [CrossRef] [PubMed]
91. Von Mutius, E.; Vercelli, D. Farm living: Effects on childhood asthma and allergy. *Nat. Rev. Immunol.* **2010**, *10*, 861–868. [CrossRef] [PubMed]
92. Poole, J.A. Farming-Associated Environmental Exposures and Atopic Diseases. *Ann. Allergy Asthma Immunol.* **2012**, *109*, 93–98. [CrossRef] [PubMed]
93. Chrischilles, E.; Ahrens, R.; Kuehl, A.; Kelly, K.; Thorne, P.; Burmeister, L.; Merchant, J. Asthma prevalence and morbidity among rural Iowa schoolchildren. *J. Allergy Clin. Immunol.* **2004**, *113*, 66–71. [CrossRef] [PubMed]

94. Wickens, K.; Lane, J.M.; Fitzharris, P.; Siebers, R.; Riley, G.; Douwes, J.; Smith, T.; Crane, J. Farm residence and exposures and the risk of allergic diseases in New Zealand children. *Allergy* **2002**, *57*, 1171–1179. [CrossRef] [PubMed]

95. Perkin, M.R.; Strachan, D.P. Which aspects of the farming lifestyle explain the inverse association with childhood allergy? *J. Allergy Clin. Immunol.* **2006**, *117*, 1374–1381. [CrossRef] [PubMed]

96. Ege, M.J.; Frei, R.; Bieli, C.; Schram-Bijkerk, D.; Waser, M.; Benz, M.R.; Weiss, G.; Nyberg, F.; van Hage, M.; Pershagen, G.; et al. Not all farming environments protect against the development of asthma and wheeze in children. *J. Allergy Clin. Immunol.* **2007**, *119*, 1140–1147. [CrossRef] [PubMed]

97. Loss, G.; Depner, M.; Ulfman, L.H.; van Neerven, R.J.J.; Hose, A.J.; Genuneit, J.; Karvonen, A.M.; Hyvärinen, A.; Kaulek, V.; Roduit, C.; et al. Consumption of unprocessed cow's milk protects infants from common respiratory infections. *J. Allergy Clin. Immunol.* **2015**, *135*, 56–62. [CrossRef] [PubMed]

98. Radon, K.; Ehrenstein, V.; Praml, G.; Nowak, D. Childhood visits to animal buildings and atopic diseases in adulthood: An age-dependent relationship. *Am. J. Ind. Med.* **2004**, *46*, 349–356. [CrossRef] [PubMed]

99. Ege, M.J.; Herzum, I.; Büchele, G.; Krauss-Etschmann, S.; Lauener, R.P.; Roponen, M.; Hyvärinen, A.; Vuitton, D.A.; Riedler, J.; Brunekreef, B.; et al. Prenatal exposure to a farm environment modifies atopic sensitization at birth. *J. Allergy Clin. Immunol.* **2008**, *122*, 407–412.e4. [CrossRef] [PubMed]

100. De Alwis, A.A.P.; Fryer, P.J. The use of direct resistance heating in the food industry. *J. Food Eng.* **1990**, *11*, 3–27. [CrossRef]

101. Duygu, B.; Ümit, G. Application of Ohmic Heating System in Meat Thawing. *Procedia Soc. Behav. Sci.* **2015**, *195*, 2822–2828. [CrossRef]

102. Guida, V.; Ferrari, G.; Pataro, G.; Chambery, A.; Di Maro, A.; Parente, A. The Effects of ohmic and conventional blanching on the nutritional, bioactive compounds and quality parameters of artichoke heads. *LWT Food Sci. Technol.* **2013**, *53*, 569–579. [CrossRef]

103. Stancl, J.; Zitny, R. Milk fouling at direct ohmic heating. *J. Food Eng.* **2010**, *99*, 437–444. [CrossRef]

104. Varghese, K.S.; Pandey, M.C.; Radhakrishna, K.; Bawa, A.S. Technology, applications and modelling of ohmic heating: A review. *J. Food Sci. Technol.* **2014**, *51*, 2304–2317. [CrossRef] [PubMed]

105. Pereira, R.N.; Vincente, A.A. Novel technologies for Milk Processing. In *Engineering Aspects of Milk and Dairy Products*; Taylor & Francis: Boca Raton, FL, USA, 2010; pp. 155–174.

106. Jaeger, H.; Roth, A.; Toepfl, S.; Holzhauser, T.; Engel, K.-H.; Knorr, D.; Vogel, R.F.; Bandick, N.; Kulling, S.; Heinz, V.; et al. Opinion on the use of ohmic heating for the treatment of foods. *Trends Food Sci. Technol.* **2016**, *55*, 84–97. [CrossRef]

107. Tucker, G. Commercially successful applications. In *Ohmic Heating in Food Processing*; CRC Press: Boca Raton, FL, USA, 2014; ISBN 978-1-4200-7108-5.

108. Cappato, L.P.; Ferreira, M.V.S.; Guimaraes, J.T.; Portela, J.B.; Costa, A.L.R.; Freitas, M.Q.; Cunha, R.L.; Oliveira, C.A.F.; Mercali, G.D.; Marzack, L.D.F.; et al. Ohmic heating in dairy processing: Relevant aspects for safety and quality. *Trends Food Sci. Technol.* **2017**, *62*, 104–112. [CrossRef]

109. Chandrasekaran, S.; Ramanathan, S.; Basak, T. Microwave food processing—A review. *Food Res. Int.* **2013**, *52*, 243–261. [CrossRef]

110. Choi, H.K.; Marth, E.H.; Vasavada, P.C. Use of microwave energy to inactive *Listeria monocytogenes* in milk. *Milchwissenschaft* **1993**, *48*, 200–203.

111. Choi, H.K.; Marth, E.H.; Vasavada, P.C. Use of microwave energy to inactive *Yersinia enterocolitica* and *Campylobacter jejuni* in milk. *Milchwissenschaft* **1993**, *48*, 134–136.

112. Galuska, P.J.; Kolarik, R.W.; Vasavada, P.C.; Marth, E.H. Inactivation of *Listeria monocytogenes* by microwave treatment. *Dairy Sci.* **1989**, *72*, 139.

113. Khalil, H.; Villota, R. Comparative study on injury and recovery of *Staphylococcus aureus* using microwaves and conventional heating. *J. Food Prot.* **1988**, *51*, 181–186. [CrossRef]

114. Lopez-Fandino, R.; Villamiel, M.; Corzo, N.; Olano, A. Assessment of the thermal-treatment of milk during continuous microwave and conventional heating. *J. Food Prot.* **1996**, *59*, 889–892. [CrossRef]

115. Merin, U.; Rosenthal, I. Pasteurisation of milk by microwave irradiation. *Milchwissenschaft* **1984**, *39*, 643–644.

116. Stearns, G.; Vasavada, P.C. Effect of microwave processing on quality of milk. *J. Food Prot.* **1986**, *49*, 853–858.

117. Villamiel, M.; López-Fandiño, R.; Corzo, N.; Martinez-Castro, I.; Olano, A. Effects of continuous-flow microwave treatment on chemical and microbiological characteristics of milk. *Z. Lebensm. Unters. Forch.* **1996**, *201*, 15–18. [CrossRef]

118. Villamiel, M.; López-Fandiño, R.; Olano, A. Microwave pasteurisation in a continuous flow unit. Shelf life of cow's milk. *Milchwissenschaft* **1996**, *51*, 674–677.

119. Ryynänen, S.; Tuorila, H.; Hyvönen, L. Perceived temperature effects on microwave heated meals and meal components. *Food Serv. Technol.* **2001**, *1*, 141–148. [CrossRef]

120. Mishra, V.K.; Ramchandran, L. Novel Thermal Methods in Dairy Processing. In *Emerging Dairy Processing Technologies: Opportunities for the Dairy Industry*; Datta, N., Tomasula, P.M., Eds.; John Wiley & Sons, Ltd.: Chichester, UK, 2015; pp. 33–70.

121. Knutson, K.M.; Marth, E.H.; Wagner, M.K. Use of microwave ovens to pasteurize milk. *J. Food Prot.* **1988**, *51*, 715–719. [CrossRef]

122. Rahman, M.S. *Handbook of Food Preservation*, 2nd ed.; CRC Press: Boca Raton, FL, USA, 2007.

123. Sierra, I.; Vidal-Valverde, C.; Olano, A. The effects of continuous flow microwave treatment and conventional heating on the nutritional value of milk as shown by influence on vitamin B1 retention. *Eur. Food Res. Technol.* **1999**, *209*, 352–354. [CrossRef]

124. Bai, Y.; Saren, G.; Huo, W. Response surface methodology (RSM) in evaluation of the vitamin C concentrations in microwave treated milk. *J. Food Sci. Technol.* **2015**, *52*, 4647–4651. [CrossRef] [PubMed]

125. Bendicho, S.; Barbosa-Cánovas, G.V.; Martín, O. Milk processing by high intensity pulsed electric fields. *Trends Food Sci. Technol.* **2002**, *13*, 195–204. [CrossRef]

126. Buckow, R.; Chandry, P.S.; Ng, S.Y.; McAuley, C.M.; Swanson, B.G. Opportunities and challenges in pulsed electric field processing of dairy products. *Int. Dairy J.* **2014**, *34*, 199–212. [CrossRef]

127. Bermúdez-Aguirre, D.; Fernández, S.; Esquivel, H.; Dunne, P.C.; Barbosa-Cánovas, G.V. Milk Processed by Pulsed Electric Fields: Evaluation of Microbial Quality, Physicochemical Characteristics and Selected Nutrients at Different Storage Conditions. *J. Food Sci.* **2011**, *76*, S289–S299. [CrossRef] [PubMed]

128. McAuley, C.M.; Singh, T.K.; Haro-Maza, J.F.; Williams, R.; Buckow, R. Microbiological and physicochemical stability of raw, pasteurised or pulsed electric field-treated milk. *Innov. Food Sci. Emerg. Technol.* **2016**, *38*, 365–373. [CrossRef]

129. Buckow, R.; Semrau, J.; Sui, Q.; Wan, J.; Knoerzer, K. Numerical evaluation of lactoperoxidase inactivation during continuous pulsed electric field processing. *Biotechnol. Prog.* **2012**, *28*, 1363–1375. [CrossRef] [PubMed]

130. Sharma, P.; Oey, I.; Bremer, P.; Everett, D.W. Microbiological and enzymatic activity of bovine whole milk treated by pulsed electric fields. *Int. J. Dairy Technol.* **2017**. [CrossRef]

131. Voigt, D.D.; Kelly, A.L.; Huppertz, T. High-Pressure Processing of Milk and Dairy Products. In *Emerging Dairy Processing Technologies—Opportunities for the Dairy Industry*; Datta, N., Tomasula, P.M., Eds.; John Wiley & Sons, Ltd.: Chichester, UK, 2015; pp. 71–92.

132. Evrendilek, G. Non-Thermal Processing of Milk and Milk Products for Microbial Safety. In *Dairy Microbiology and Biochemistry*; CRC Press: Boca Raton, FL, USA, 2014; pp. 322–355.

133. Patterson, M.F. Microbiology of pressure-treated foods. *J. Appl. Microbiol.* **2005**, *98*, 1400–1409. [CrossRef] [PubMed]

134. Villamiel, M.; Schutyser, M.A.I.; De Jong, P. Novel Methods of Milk Processing. In *Milk Processing and Quality Management*; Tamine, A.Y., Ed.; Wiley-Blackwell: Oxford, UK, 2009; pp. 205–236.

135. Huppertz, T.; Fox, P.F.; Kelly, A.L. High pressure-induced denaturation of alpha-lactalbumin and beta-lactoglobulin in bovine milk and whey: A possible mechanism. *J. Dairy Res.* **2004**, *71*, 489–495. [CrossRef] [PubMed]

136. Tomasula, P.M.; Bonnaillie, L.M. Crossflow Microfiltration in the Dairy Industry. In *Emerging Dairy Processing Technologies—Opportunities for the Dairy Industry*; Datta, N., Tomasula, P.M., Eds.; John Wiley & Sons, Ltd.: Chichester, UK, 2015; pp. 1–31.

137. Zisu, B.; Chandrapala, J. High Power Ultrasound Processing in Milk and Dairy Products. In *Emerging Dairy Processing Technologies—Opportunities for the Dairy Industry*; Datta, N., Tomasula, P.M., Eds.; John Wiley & Sons, Ltd.: Chichester, UK, 2015; pp. 149–180.
138. Villamiel, M.; de Jong, P. Influence of high-intensity ultrasound and heat treatment in continuous flow on fat, proteins and native enzymes of milk. *J. Agric. Food Chem.* **2000**, *48*, 472–478. [CrossRef] [PubMed]

beverages

MDPI

Short Note

Milk and Its Sugar-Lactose: A Picture of Evaluation Methodologies

Loretta Gambelli

Council for Agriculture Research and Analysis of the Agrarian Economy, Research Center CREA-Food and Nutrition, Via Ardeatina 546, 00178 Rome, Italy; loretta.gambelli@crea.gov.it; Tel.: +39-065-149-4411

Academic Editor: Alessandra Durazzo
Received: 30 May 2017; Accepted: 26 June 2017; Published: 13 July 2017

Abstract: Lactose is the major disaccharide found in milk, and is catabolized into glucose and galactose by the enzyme lactase. Lactose is an important energy source and ssometimes it is referred to simply as milk sugar, as it is present in high percentages in dairy products. Lactose is the primary source of carbohydrates during mammal development, and represents 40% of the energy consumed during the nursing period. Lactose-intolerance individuals have a lactase deficiency; therefore, lactose is not completely catabolized. Lactose intolerance is a significant factor in the choice of diet for many sick people, therefore its content in foods must be monitored to avoid disorders and illnesses. This has created the need to develop simple methods, such as polarimetry, gravimetric, middle infrared, differential pH and enzymatic monitoring, but all these methods are time-consuming, because they required extensive sample preparation and cannot differentiate individual sugars. In order to quantify low levels of lactose, new and more accurate analytical methods have been developed. Generally, they require equipment such as HPLC or High Performance Anion Exchange with Pulsed Amperometric Detection (HPAE-PAD).

Keywords: lactose; analytical methodologies; nutrition

1. Lactose and Its Peculiar Role Nutrition

Milk and dairy products represent a food category of absolute centrality in the diet of the Italian population. The Italian Food Pyramid has entered these products at an intermediate level, attributing to milk, yogurt and cheese a consumption mode in the order of 2–3 servings per day.

Milk is a complex fluid foodstuff secreted by the breast glands of female mammals. It is a complete food, since it contains nearly all the necessary nutrients to sustain the life and growth of a newborn.

Even if different types of milk can be used for human consumption, such as sheep, goat, donkey and other infant formula milks, in addition to breast milk, in the first period of life, usually when we speak of milk we are referring to cow's milk [1].

Among the nutrients contained in milk, saccharides play a key role. They are present in amounts of up to ca. 10%, depending on the mammal [2], and they are represented almost exclusively by lactose (98% of the sugars present in milk), which is not found in any other food, and is important for the development of the nervous tissue in the first few months of life. The main role of lactose is to provide the newborn with galactose for the synthesis of the nerve structures (myelin sheaths).

D-lactose is present in milk in two anomeric forms, α-lactose and β-lactose (ratio 2:3), depending on the pyranose form (α or β) of glucose. Galactose is always present in the β-pyranose form.

These anomers also have different physical properties, such as melting point and, primarily, solubility in water, as β-lactose is much more soluble than α-lactose [3,4].

Lactose is synthesized in the mammary tissue from the conversion of a part of the glucose present in the blood to galactose. This synthesis involves the complex lactose synthetase, which is provided by the galactosyltransferase and α-lactalbumin [5].

2. Methodologies for the Determination of Lactose

Detection of lactose is very important because of lactose intolerance diseases. Small quantities of it are present in many foods—not only in milk and dairy products—and therefore, traces of lactose may be present in food under the following headings: milk solids, whey, curds, skim milk powder, and skim milk solids, meaning that lactose is present, and must be reported on the food label. Lactose intolerance is a significant factor in the choice of the diet of people sensitive or intolerant to this sugar, so its content in foods must be monitored to avoid disorders and diseases [6,7]. It is therefore important to quantify lactose with precision and accuracy in these products. The chosen method should be one that is economical, rapid, and sensitive.

A considerable number of methods for determining carbohydrates in milk have been developed, including older and less sensitive ones, such as gravimetric, polarimetric, enzymatic, or spectrophotometric analysis, as well as more specific and sensible methods such as HPLC-RI and HPAEC-PAD.

The most common methods used to quantify lactose in milk are described below:

2.1. Polarimetric

Polarimetric methods are based on the measurement of the specific rotation of polarized light by chiral molecules such as lactose in a skimmed and deproteinized milk filtrate [8,9].

2.2. Gravimetric

Gravimetric methodologies are founded on the decrease of copper sulfate to cuprous oxide precipitated by the addition of potassium hydroxide in the presence of aldoses and ketoses. The lactose content is calculated after weighing the cuprous oxide formed, by using empirical tables that allow the conversion of the cuprous oxide formed in terms of lactose [8,10].

2.3. Mid-Infrared

Infrared Spectroscopic Methods are based on the absorbance of infrared energy by the hydroxyl groups (OH) of lactose molecules. Lactose determination in mid-infrared (MIR) spectroscopy is carried out at 1042 cm^{-1} [11]. Early instruments were entirely filter-based, using pairs (sample and reference) of optical filters to select a band of wavelengths for the measurement of fat, protein, and lactose.

Now, more recent instruments utilize an interferometer to acquire complete spectrum information within the MIR region using Fourier Transform Infrared Spectroscopy (FTIR); in this way, it is possible to obtain an extensive computing and data manipulation capabilities [12–15].

2.4. Enzymatic

A large number of enzymatic methodologies able to quantify lactose have been reported [16–19]. They have been characterized by the common reaction of enzymatic hydrolysis of lactose to glucose and galactose, followed by the enzymatic determination of one of the liberated monosaccharides. The amount of lactose in the sample is given by the difference between monosaccharide content before and after hydrolysis.

2.4.1. NAD Enzymatic

In my opinion, the most used enzymatic method to measure galactose is based on its oxidation by β-galactose dehydrogenase to galacturonic acid in the presence of nicotinamide-adenine dinucleotide (NAD) that is reduced to NADH, as reported by the following reaction:

β-D-Galactose + NAD⁺ β-D-Galactose Dehydrogenase → D-Galactono-γ-Lactone + NADH + H⁺

Absorbance of NADH at 340 nm is calculated as the difference between the readings before and after the addition of the enzyme, galactose dehydrogenase [20–22]. Although this UV method is specific and accurate, as the measurements of NADH require reading in the UV range, replacement of NAD by thio-NAD and measurement in the visible range at 405 or 415 nm can also be done. This variation allows the simultaneous quantification of D-galactose concentrations in several samples using microplate-readers, rather than UV spectrophotometers [16–19].

2.4.2. Differential pH

The differential pH technique for determining of lactose and lactulose in milk samples is based on changes in the pH owing to enzymatic reactions. Lactose determination is performed by measuring the pH change caused by the reaction of glucose and ATP in the presence of hexokinase (HK) before and after treatment of the sample with β-galactosidase. The determination of lactulose is carried out by treating the sample with a mixture of β-galactosidase and glucokinase in the presence of ATP. After 3 h, the pH change is measured, HK is added, and the pH change is monitored for 4 h to observe the d-fructose-6-phosphate formation [23].

2.5. HPLC RI/HPAEC-PAD

HPLC remains as one of the most extensively used techniques. It has been widely used for separating a large variety of carbohydrates, especially in foods, as it is particularly advantageous in terms of the speed and simplicity of sample preparation. HPLC allows direct detection of carbohydrates, as they can absorb low UV wavelengths.

However, detection in this spectrum area (below 200 nm) is difficult due to its low sensitivity and selectivity; it also requires the use of high-quality and expensive reagents. The most common sugar-detection system after HPLC separation is the refractive index; however, the response of this detector is very poor and non-specific, quite sensitive to changes in temperature, pressure, and solvent composition, and does not allow gradients. If a refractive index (RI) is used for detection, the analysis is straightforward, but not very sensitive, with a Limit of Detection (LOD) of 250 mg/L and a Limit of Quantification (LOQ) of 380 mg/L having been reported [24].

Several chromatographic methods are available for the separation of carbohydrates, reverse phase systems and cation exchange are the most widely used [25–29].

The separation in reverse-phase partition chromatography is based on the principle of hydrophobic interactions derived from the repulsive forces among relatively polar solvents, nonpolar analytes, and nonpolar stationary phases. Alkylated and aminoalkylated silica gels are most frequently used as stationary phases, in combination with aqueous methanol or aqueous acetonitrile as mobile phases, and the separation is carried out by hydrophobic and polar interactions and partition [24,30,31].

Traditional adsorption chromatography was almost universally replaced by Ion-Exchange Chromatography (IEC). Carbohydrates are separated by charge differences using two types of ion-exchanger—anionic and cationic—where the compounds are negatively and positively charged, respectively [32,33].

Several methodologies based on cationic-exchange HPLC chromatography have been optimized to quantify carbohydrates in a lot of dairy products, using different stationary and mobile phases, such as Amine with calcium as counterion, and Sugar Pak [27–29].

A second, more sensitive method is High-Performance Anion-Exchange Chromatography (HPAEC) with Pulsed Amperometric Detection (PAD). High-performance anion-exchange chromatography (HPAEC) coupled with PAD is an alternative analytical technique that presents high sensitivity and good resolution compared with non-derivatized carbohydrates.

Carbohydrate separation and elution order is based on the differences in their pKa values, in fact HPAEC takes advantage of the weakly acidic nature of carbohydrates to give highly selective separations at high pH using a strong anion exchange stationary phase. Columns are packed with poly(styrene-divinylbenzene)-based stationary phases functionalized with alkyl quaternary ammonium groups [34]. The possible co-elution of closely related carbohydrates with very similar retention times represents the main problem of HPAEC-PAD. Cataldi et al. [35] optimized a rapid and sensitive HPAEC-PAD method using 10–12 mM NaOH modified with 1–2 mM barium acetate in heated milks; in this way, the separation and quantification of lactulose and lactose along with other carbohydrates was obtained in milk and milk products [35,36].

In one recent application note, 248, the Dionex Corporation, now part of Thermo Fisher, suggested using this technique to quantify lactose in lactose-free products, and indicated a LOD lower than 1 mg/L [37,38].

Obviously, all the methods above describe present advantages and disadvantages, but in my opinion this last method presents a good cost/benefit ratio.

3. Conclusions

Lactose, a very important nutrient during the neonatal years, is not always tolerated in adults. In this short review, various methods, from the oldest to the more modern and innovative, for determining lactose in milk have been reported. Considering how widespread lactose intolerance is, today lactose analysis should be considered as routine analysis. In order to choose the most suitable method, important parameters such as sensitivity, precision, accuracy, and speed of analysis have been taken under consideration.

But unfortunately, this is not always possible, because the high cost of scientific instruments, reagents, laboratories and the adequate training of technicians in some situations, for example, in undeveloped countries, makes it difficult to perform sophisticated analysis. However, under working conditions, each operator must choose the best methods available. That is the principal aim of this review.

Conflicts of Interest: The authors declare no conflict of interest.

References

1. Nickerson, T.A. Lactose. In *Fundamentals in Dairy Chemistry*, 2nd ed.; Webb, B.H., Johnson, A.H., Alford, J.A., Eds.; Avi Publishing Co.: Westport, CT, USA, 1974; Chapter 6.
2. Fox, P.F.; Mcsweeney, P.L.H. *Dairy Chemistry and Biochemistry*; Blackie Academice Professional: London, UK, 1998.
3. Ganzle, M.G. Enzymatic synthesis of galacto-oligosaccharides and other lactose derivatives (hetero-oligosaccharides) from lactose. *Int. Dairy J.* **2012**, *22*, 116–122. [CrossRef]
4. Idda, I.; Spano, N.; Ciulu, M.; Nurchi, V.M.; Panzanelli, A.; Pilo, M.I.; Sanna, G. Gas chromatography analysis of major free mono- and disaccharides in milk: Method assessment, validation, and application to real samples. *J. Sep. Sci.* **2016**, *39*, 4577–4584. [CrossRef] [PubMed]
5. Harju, M.; Kallioinen, H.; Tossavainen, O. Lactose hydrolysis and other conversions in dairy products: Technological aspects. *Int. Dairy J.* **2012**, *22*, 104–109. [CrossRef]
6. EFSA NDA Panel (EFSA Panel on Dietetic Products, Nutrition and Allergies). Scientific opinion on lactose thresholds in lactose intolerance and galactosaemia. *EFSA J.* **2010**, *8*, 29. [CrossRef]
7. Brown-Esters, O.; Namara, M.P.; Savaiano, D. Dietary and biological factors influencing lactose intolerance. *Int. Dairy J.* **2012**, *22*, 98–103. [CrossRef]
8. Lactose. In *Official Methods of Analysis*, 15th ed.; Helrich, K., Ed.; Association of Official Analytical Chemists (AOAC): Washington, DC, USA, 1990; p. 810.
9. Official Method 896.01. Lactose in Milk. Polarimetric Method. In *Official Methods of Analysis of AOAC International*; AOAC: Washington, DC, USA, 2005.
10. Official Method 930.28. Lactose in Milk. Gravimetric Method. In *Official Methods of Analysis of AOAC International*; AOAC: Washington, DC, USA, 2005.
11. Lactose. Infrared Milk Analysis. In *Official Methods of Analysis*, 18th ed.; Horwitz, W., Ed.; AOAC: Washington, DC, USA, 2005.
12. Linch, J.M.; Barbano, D.M.; Schweisthal, M.; Fleming, J.R. Precalibration evaluation procedures for mid-infrared milk analyzers. *J. Dairy Sci.* **2006**, *89*, 2761. [CrossRef]

13. Official Method 972.16 Lactose in Milk. Mid-Infrared Method. In *Official Methods of Analysis of AOAC International*; AOAC: Washington, DC, USA, 2005.
14. IDF. *International Standard 141C. Whole Milk-Determination of Milkfat, Protein and Lactose Content, Guidance on the Operation of Mid-Infrared Instruments*; Int. Dairy Fed.: Brussels, Belgium, 2000.
15. Kittivachra, R.; Sanguandeekul, R.; Sakulbumrungsil, R.; Phongphanphanee, P.; Srisomboon, J. Determination of essential nutrients in raw milk. *Songklanakarin J. Sci. Technol.* **2006**, *28*, 115–120.
16. Shapiro, F.; Shamay, A.; Silanikove, N. Determination of lactose and D-galactose using thio-NAD+ instead of NAD+. *Int. Dairy J.* **2002**, *12*, 667. [CrossRef]
17. *ISO 5765-1:2002. Dried Ice-Mixes and Processed Cheese—Determination of Lactose Content—Part 1: Enzymatic Method Utilizing the Glucose Moiety of the Lactose*; International Organization for Standardization: Geneva, Switzerland, 2002.
18. *ISO 5765-2:2002. Dried Ice-Mixes and Processed Cheese—Determination of Lactose Content—Part 2: Enzymatic Method Utilizing the Galactose Moiety of the Lactose*; International Organization for Standardization: Geneva, Switzerland, 2002.
19. Official Method 984.15. Lactose in Milk. Enzymatic Method. In *Official Methods of Analysis of AOAC International*; AOAC: Washington, DC, USA, 2005.
20. Coffey, R.G.; Reithel, F.J. An enzymic determination of lactose. *Anal. Biochem.* **1969**, *32*, 229. [CrossRef]
21. Lynch, J.M.; Barbano, D. Determination of the lactose content of fluid milk by spectrophotometric enzymatic analysis using weight additions and path length adjustment: Collaborative study. *J. AOAC Intern.* **2007**, *90*, 196.
22. Kleyn, D.H. Determination of lactose by an enzymatic method. *J. Dairy Sci.* **1985**, *68*, 2791–2798. [CrossRef]
23. *Milk: Determination of Lactose Content. Enzimatic Method Using Difference in pH. ISO Standard 26462*; International Organization of Standardization: Geneva, Switzerland, 2010.
24. Chavez-Servin, J.L.; Castellote, A.I.; Lopez-Sabater, M.C. Analysis of mono-and disaccharides in milk-based formulae by high-performance liquid chromatography with refractive index detection. *J. Chromatogr. A* **2004**, *1043*, 211–215. [CrossRef] [PubMed]
25. Richmond, M.L.; Barfuss, D.L.; Harte, B.R.; Gray, J.I.; Stine, C.M. Separation of carbohydrates in dairy products by high performance liquid chromatography. *J. Dairy Sci.* **1982**, *65*, 1394. [CrossRef]
26. West, L.G.; Llorente, M.A. High performance liquid chromatographic determination of lactose in milk. *J. Assoc. Off. Anal. Chem.* **1981**, *64*, 805. [PubMed]
27. Pirisino, J.F. High performance liquid chromatographic determination of lactose, glucose and galactose in lactose-reduced milk. *J. Food Sci.* **1983**, *48*, 742. [CrossRef]
28. Mullin, W.J.; Emmons, D.B. Determination of organic acids and sugars in cheese, milk and whey by high performance liquid chromatography. *Food Res. Int.* **1997**, *30*, 147. [CrossRef]
29. Elliot, J.; Dhakal, A.; Datta, N.; Deeth, H.C. Heat-induced changes in UHT milks. Part 1. *Aust. J. Dairy Technol.* **2003**, *58*, 3.
30. Honda, S. High-performance liquid chromatography of mono- and oligosaccharides. *Anal. Chem.* **1984**, *140*, 1. [CrossRef]
31. Manzi, P.; Pizzoferrato, L. HPLC determination of lactulose in heat treated milk. *Food Bioprocess Technol.* **2013**, *6*, 851–857. [CrossRef]
32. Kennedy, J.F.; Pagliuca, G. Oligosaccharides. In *Carbohydrate Analysis. A Practical Approach*; (The Practical Approach Series); Chaplin, M.F., Kennedy, J.F., Eds.; Oxford University Press: New York, NY, USA, 1994; Chapter 2.
33. Macrae, R. Applications of HPLC to food analysis. In *HPLC in Food Analysis*; Macrae, R., Ed.; Academic Press: San Diego, CA, USA, 1988; Chapter 2.
34. Cataldi, T.R.I.; Campa, C.; De Benedetto, G.E. Carbohydrate analysis by high performance anion-exchange chromatography with pulsed amperometric detection: The potential is still growing. *Fresen. J. Anal. Chem.* **2000**, *368*, 739. [CrossRef]
35. Cataldi, T.R.I.; Angelotti, M.; Bufo, S.A. Method development for the quantitative determinationof lactulose in heat-treated milks by HPAEC-with pulsed amperometric detection. *Anal. Chem.* **1999**, *71*, 4919. [CrossRef] [PubMed]
36. Cataldi, T.R.I.; Angelotti, M.; Bianco, G. Determination of mono-and disaccharides in milkand milk products by high performance anion-exchange chromatography with pulsed amperometric detection. *Anal. Chim. Acta* **2003**, *485*, 43. [CrossRef]

37. Monti, L.; Negri, S.; Meucci, A.; Stroppa, A.; Galli, A.; Contarini, G. Lactose, galactose and glucose determination in naturally "lactose free" hard cheese: HPAEC-PAD method validation. *Food Chem.* **2017**, *220*, 18–24. [CrossRef] [PubMed]

38. Determinatio of Lactose in Lactose-Free Milk products by High-Performance Anion-Exchange Chromatography with Pulsed Amperometric Detection. Available online: https://tools.thermofisher.com/content/sfs/brochures/AN-248-Lactose-Milk-Products-HPAE-PAD-AN70236-EN.pdf (accessed on 13 July 2017).

beverages

MDPI

Review

Therapeutic Potential of Milk Whey

Charu Gupta * and Dhan Prakash

Amity Institute for Herbal Research and Studies, Amity University Uttar Pradesh,
Sector-125, Noida-201313, India; dprakash_in@yahoo.com
* Correspondence: charumicro@gmail.com; Tel.: +91-120-439-2549

Academic Editor: Alessandra Durazzo
Received: 25 April 2017; Accepted: 27 June 2017; Published: 5 July 2017

Abstract: Milk whey—commonly known as cheese whey—is a by-product of cheese or casein in the dairy industry and contains usually high levels of lactose, low levels of nitrogenous compounds, protein, salts, lactic acid and small amounts of vitamins and minerals. Milk whey contains several unique components like immunoglobulins (Igs), lactoferrin (Lf), lactoperoxidase (Lp), glycomacropeptide (GMP) and sphingolipids that possess some important antimicrobial and antiviral properties. Some whey components possess anticancer properties such as sphingomyelin, which have the potential to inhibit colon cancer. Immunoglobulin-G (IgGs), Lp and Lf concentrated from whey participates in host immunity. IgGs binds with bacterial toxins and lowers the bacterial load in the large bowel. There are some whey-derived carbohydrate components that possess prebiotic activity. Lactose support lactic acid bacteria (such as *Bifidobacteria* and *Lactobacilli*). Stallic acids, an oligosaccharide in whey, are typically attached to proteins, and possess prebiotic properties. The uniqueness of whey proteins is due to their ability to boost the level of glutathione (GSH) in various tissues and also to optimize various processes of the immune system. The role of GSH is very critical as it protects the cells against free radical damage, infections, toxins, pollution and UV exposure. Overall GSH acts as a centerpiece of the body's antioxidant defense system. It has been widely observed that individuals suffering from cancer, HIV, chronic fatigue syndrome and many other immune-compromising conditions have very poor levels of glutathione. The sulphur-containing amino-acids (cysteine and methionine) are also found in high levels in whey protein. Thus, the present review will focus on the therapeutic potential of milk whey such as antibiotic, anti-cancer, anti-toxin, immune-enhancer, prebiotic property etc.

Keywords: milk whey; therapeutic potential; cheese whey; antimicrobial; immune-enhancer; bioactive peptides

1. Introduction

Whey is generally released as a by-product during cheese manufacturing. The typical composition of milk comprises about 3.6% protein, out of which casein predominates to around 80% and rest 20% are called as the whey proteins. Whey proteins are unique as they contain all the essential amino acids of good quality protein. Milk whey and whey proteins have different biological and functional properties. Consequently, milk whey proteins are used in the manufacture of various products such as infant foods, nutritional products for athletes, tailor made specialized products for controlling obesity, mood control and other clinical protein supplements such as for enteral disturbances. (Yalcin, 2006 [1]).

Bioactive peptides present in milk whey are one of the most studied compounds in different dairy products. Bioactive peptides from dairy sources are majorly classified on the basis of their biological roles as anti-hypertensive, anti-oxidative, immmuno-modulant, anti-mutagenic, anti-microbial, opioid, anti-thrombotic, anti-obesity and mineral-binding agents. These bioactive peptides are produced by enzymatic hydrolysis during fermentation and gastrointestinal digestion. Thus, fermented dairy products like yogurt, cheese and buttermilk are gaining popularity worldwide and are considered as an excellent source of dairy peptides. Furthermore, these dairy products are also associated with lower risks of hypertension, coagulopathy, stroke and cancer insurgences (Sultan et al., 2014 [2]).

2. Historical Background

Historically, whey was considered as a cure for all common ailments ranging from gastrointestinal complaints to joint and ligament problems.

Nanna Rognvaldardottir, a food expert from Iceland, described whey (also called syra in Iceland), as fermented liquid that is stored in barrels.

Before its consumption, Syra is diluted with water. It is also used as a marinade or preservative for meat and other food.

Syra was the most common beverage of Icelandic people and replaced ale due to the lack of grains in that region (Rognvaldardottir, 2001 [3]).

3. Milk Whey Components

The major components of milk whey of nutraceutical potential includes beta-lactoglobulin, alpha-lactalbumin, bovine serum albumin (BSA), lactoferrin (Lf), immunoglobulins (Igs), lactoperoxidase (Lp) enzymes, glycomacropeptides (GMP), lactose and minerals. The composition of the whey liquid also depends upon the source of milk e.g., whey derived from buttermilk contains more lipid sphingomyelin as compared to those derived from cheese. Whey is a popular dietary protein supplement that is well known to possess antimicrobial activity, immune modulation, improved muscle strength and body composition. Furthermore, whey is known to prevent cardiovascular disease and osteoporosis.

The nutrient content of milk whey types (sweet whey and whey permeate) is described in Appedix Table A1 and the whey characteristics are highlighted in Appedix Table A2.

3.1. Beta-Lactoglobulin

Beta-lactoglobulin comprises approximately half of the total protein content in bovine whey, while it is absent in human milk. It is a source of both essential and branched chain amino acids. A retinol binding protein (a carrier of small hydrophobic molecules including retinoic acid) is present within the beta-lactoglobulin structure. This protein, has an ability to modulate lymphatic responses (Gupta et al., 2012; Yolken et al., 1985 [4,5]). It also has the property to bind to hydrophobic ligands such as fatty acids.

Recently, Le Maux et al. demonstrated that beta-lactoglobulin acts as a carrier molecule that alters the bio-accessibility of linoleate and linoleic acid (Le Maux et al., 2012 [6]). It also provides resistance against gastric- and simulated duodenal digestions. It also serves as a potential carrier for delivering gastric labile hydrophobic drugs. Thus it has a great potential to serve as a realistic protein candidate for safe delivery and protection of pH-sensitive drugs in the stomach (Mehraban et al., 2013 [7]).

3.2. Alpha-Lactalbumin

Alpha-lactalbumin is reported to be the second most important protein in whey quantitatively and it represents approx. 20% (*w/w*) of the total whey protein. It is completely synthesized in the mammary gland. It has significant anti-proliferative effects in human adenocarcinoma cell lines such as Caco-2 and HT-29 (Brück et al., 2014 [8]). It also kills tumour cells and has bactericidal effects in the upper respiratory systems and protective effects on gastric mucosa.

Alpha-lactalbumin plays a vital role in reducing the risk of some cancers as it constrains cell division (Ganjam et al., 1997 [9]). In yet another study, it has also been found effective in the treatment of cognitive declination. This is due to the high tryptophan content in alpha-lactalbumin, that increases the plasma tryptophan-large neutral amino acids ration (Markus et al., 2002 [10]).

3.3. Bovine Serum Albumin (BSA)

Bovine serum albumin (BSA) is not produced in the mammary gland, but is secreted in milk after its passive leakage from the blood stream. The most important property of bovine serum albumin is its ability to bind to various ligands reversibly. It is the principal carrier of fatty acids and can bind to free fatty acids and other lipids as well as flavoring compounds (Huang et al., 2004 [11]). However, this property is impaired upon denaturation on heating. BSA also inhibits tumor growth due to the modulation of activities of the autocrine growth regulatory factors (Laursen et al., 1990 [12]).

BSA also binds to fatty acids stored in the human body; and thus it also participates in lipid synthesis (Choi et al., 2002 [13]). In addition, BSA possess antioxidant activities (Tong et al., 2000 [14]).

BSA is a source of all major essential amino acids, whose therapeutic potential is still largely unexplored.

3.4. Lactoferrin (Lf)

Lactoferrin is an iron-binding glycoprotein that is present in whey fraction of milk and colostrum. It is a non-enzymatic antioxidant and consists of approximately 689 amino acid residues (Pierce et al., 1991 [15]). The concentration of lactoferrin in human milk and colostrum is reported to be about 2 mg/mL and 7 mg/mL, respectively, whereas in bovine milk and colostrum it is around 0.2 mg/mL and 1.5 mg/mL, respectively (Levay et al., 1995 [16]). Lactoferrin is a pre-dominant component of whey protein in human breast milk.

Lactoferrin is also known to be an important host defense molecule and performs a range of physiological functions such as antimicrobial, antiviral, immuno-modulatory and antioxidant activity. It has been shown in several scientific studies that lactoferrin, if administered orally, exerts several beneficial health effects on humans and animals, including anti-infective, anti-cancer and anti-inflammatory effects. Thus, lactoferrin possess the potential to be used as a food additive (El-Loly and Mahfouz, 2011 [17]).

3.5. Immunoglobulins (Igs)

Immunoglobulins (Ig) are antibodies and chemically gamma-globulins. The whey fraction of milk contains a major amount of immunoglobulins, comprising approximately 10%–15% of total whey proteins. Numerous studies have proved the therapeutic potential of immunoglobulins. They are known to possess vital biological properties.

In an *in vitro* study, it was shown that bovine milk-derived IgG suppresses human lymphocyte proliferative response to T cells at concentrations as low as 0.3 mg/mL of IgG. It was further concluded that bovine milk IgG concentration varies between 0.6–0.9 mg/mL and therefore confer immunity that is transmitted to humans (El-Loly, 2007 [18]). Previous studies have shown that unpasteurized cow milk contain specific antibodies to human rotavirus, and against bacteria such as *E. coli, Salmonella enteriditis, S. typhimurium*, and *Shigella flexneri* (Losso et al., 1993 [19]).

Nowadays, commercial colostral whey-derived immunoglobulin preparations are widely available in the market that are marketed under the category of feed supplements and newborn farm animal substitutes (Scammell, 2001 [20]).

There are few studies that have reported that the efficacy of these non-specific immunoglobulin products in the prevention and treatment of gastrointestinal infections is variable in different animal studies (Garry et al., 1996 [21]).

However, Hilpert et al. [22] described the production of 'hyper-immune milk' by immunizing the cows with specific antibodies and to prepare a whey protein concentrate for protecting young animals from disease.

An increasing number of controlled clinical studies have shown that the oral administration of immune milk preparations containing high titers of specific antibodies can provide effective protection and, to some extent, may also have therapeutic value against gastrointestinal infections in humans (Korhonen et al., 2000; Lilius and Marnila, 2001 [23,24]).

3.6. Lactoperoxidase (LPO) Enzymes

Milk whey contains many types of enzymes, such as lactoperoxidase, hydrolases, transferases, lyases, proteases and lipases. Lactoperoxidase accounts for 0.25–0.5 percent of total protein found in whey. Lactoperoxidase is an important enzyme present in the whey fraction of milk. It is the most abundant enzyme in whey following the curding process. It has the property to catalyze certain molecules, and reduction of hydrogen peroxide. This enzyme system catalyzes peroxidation of thiocyanate and some halides (such as iodine and bromium), which ultimately generates products that inhibit and/or kill a range of bacterial species. It is heat tolerant and therefore lactoperoxidase is not inactivated during the pasteurization process, thereby suggesting its stability as a preservative.

The biological significance of this enzyme is that it has a natural protection system against the invasion of microorganisms. Besides its antiviral effect, it protects animal cells against various damages and peroxidative effects (de Wit and Van Hooydonk, 1996 [25]).

Lactoperoxidase also provides defense system against pathogen microorganisms from the digestive system of neonatal babies. The LPO enzyme functions as a non-immune biological defense system of mammals and catalyzes the oxidation of the thiocyanate ion into the antibacterial hypothiocyanate (Reiter and Perraudin, 1991 [26]). In yet another study, the oral administration of lactoperoxidase attenuated pneumonia in influenza virus-infected mice suppressed the infiltration of inflammatory cells in the lungs (Shin et al., 2005 [27]).

Thus the major application of lactoperoxidase is a protective factor against infectious microbes.

3.7. Glycomacropeptides (GMP)

Glycomacropeptides are also known as Casein Macro-Peptide (CMP) and constitutes about 10%–15% of protein in whey. It is produced in whey due to the breakdown of casein by chymosin enzyme during cheese-making process. It contains large numbers of branched chain amino acids; but lacks the aromatic amino acids such as phenylalanine, tryptophan and tyrosine.

The other advantage is that it can be safely administered to patients suffering with phenylketonuria (PKU) as it lacks phenylalanine (Marshall, 2004 [28]).

The GMP inhibits gastric acid secretions and modifies the blood concentration of regulatory digestive peptides. It induces satiety as it induces the release of cholecystokinin but similar results were not observed in human-fed GMP (Gustafson et al., 2001 [29]). The other role of GMP is to inhibit the adhesion of cariogenic bacteria such as *Streptococcus mutans*, *Sanguis* and *Sobrinus* to oral surfaces and therefore can modify the composition of plaque bacteria to control its acid production and, in turn, reduce the demineralization of enamel and promote re-mineralization. The GMP is also a source of N-acetyl-necromatic acid and its dietary intake can increase the sialic acid content of saliva that effects its viscosity and protective function (Gupta et al., 2012 [4]).

3.8. Lysozyme (N-Acetylmuramide Glycan-Hydrolase)

Lysozyme is a hydrolytic enzyme that is widely distributed in nature and occurs in many body fluids and tissues of living organisms (Fox and Kelly, 2006 [30]). Although the highest concentration of the enzyme is found in tears and egg white protein, but is also present in human milk. It exhibits antibacterial activity only against gram-positive bacteria.

This enzyme has wider industrial applications as a food additive, in medical diagnostics, pharmacology and veterinary medicine. It is also used in the treatment of bacterial and viral infections, skin and eye diseases, periodontitis, leukemia and cancer (Lesnierowsk, 2009; Benkerroum, 2008 [31,32]).

3.9. Proteose Peptone Component 3 (PP₃)

These are those proteins that are left in solution after the milk is heated at high temperature followed by its acidification up to pH 4.7.

The proteose peptone 3 component is found only in whey excluding from the human source. It is produced during the fermentation of fat-free bovine milk and enhances the production of monoclonal antibodies by human hybridoma cells (Krissansen, 2007 [33]).

4. Therapeutic Properties of Milk Whey

4.1. Antimicrobial

Milk whey—also known as cheese whey (Lactoserum)—contains several unique components that possess some important antimicrobial and antiviral properties. These are immunoglobulins, lactoferrin, lactoperoxidase, glycomacropeptide and sphingolopids. All these compounds are able to survive after their passage through stomach and small intestine and exert their biological effects in the large intestine. The important whey components that provide antimicrobial action in the intestinal tract are the immunoglobulins like IgG, IgM and IgA. IgG binds to toxin produced by *Clostridium difficile*, thereby reducing diarrhoea, dehydration and muscle aches. GMP inhibits binding of cholera toxin to receptors in the intestinal tract.

Lactoferrin, an iron binding protein present in milk whey, possesses antibacterial properties. It sequesters iron from bacteria (Troost et al., 2001 [34]). Since pathogens in particular have high iron requirements for metabolism and growth, this property of lactoferrin makes it broadly antimicrobial in nature. Lactobacilli can utilize lactoferrin-bound iron, thus allowing lactoferrin to both inhibit pathogenic bacteria and support the growth of lactobacilli.

Moreover, lactoferrin possess important antiviral activity as.it directly interacts with selected viral pathogens, inhibits virus replication and their ability to attach to colonic epithelial cells. Viral inhibition also results through immune modulation benefits of lactoferrin.

Another whey-derived important protein component is lactoperoxidase (Lp) with potent antimicrobial properties. It catalyzes the oxidation of thiocyanate into hypothiocynate ion, which is a strong oxidizing agent. Hypothiocynate ions disrupts bacterial cell membranes. This "Lactoperoxidasesystem" is therefore used in a milk-preservation system. Whey-derived phospholipids such as sphingolipids are metabolized in the gastrointestinal (GI) tract and produce sphingosine and lysosphingomyelin. These compounds possess powerful bactericidal agents in vitro.

Cheese whey (Lactoserum) was used as an inexpensive medium to produce bacteriocin by *Bacillus* sp. P11 (Leães et al., 2010 [35]). Bacteriocins are antimicrobial peptides produced by lactic acid bacteria group.

The lactoferrin, α-Lactalbumin and β-Lactglobulin were also studied for their antagonistic activity against Human Immunodeficiency Virus type-1 (HIV-1) (Chatterton et al., 2006 [36]). In particular, β-Lactglobulin is potential agent for preventing the transmission of genital herpes virus infections and the spread of HIV. The lactoferrin and lactoferricin exhibit inhibitory activity against a broad range of microorganisms including Gram-negativeand, Gram-positive bacteria, yeast, fungi and parasitic protozoa (Takakura et al., 2003 [37]). They also inhibit the growth of food-borne pathogens such as *E. coli* and *Listeria monocytogenes* (Floris et al., 2003 [38]).

The Lf exhibits antiviral property against HIV, Human Cytomegalovirus (HCMV), herpes viruses, Human Papilloma Virus (HPV), alpha-virus and hepatitis C, B and G viruses. Overall, whey proteins activate immune cell and/or prevent infection. They show promise to help combat rota-viral diarrhea, which is a common infection in children (Wolber et al., 2005 [39]). The natural antimicrobial action of Lp is used in a range of oral healthcare products and in the prevention and treatment of xerostimia (dry mouth). Whey protein concentrate supplementation can thus decrease the occurrence of associated co-infections (Solak and Akin, 2012 [40]).

4.2. Anti-Cancerous

Some cheese whey components possess anticancer properties. The sulphur-containing amino-acids including cysteine and methionine, are found in good amounts in whey protein. Cysteine and methionine are utilized in glutathione synthesis. Glutathione is a substrate for two classes of enzymes that catalyze detoxification compounds and bind mutagens and carcinogens, thus facilitating their elimination from the body. The ability of lactoferrinto bind iron is another added advantage in the prevention of colon cancer. They induce apoptosis in tumour cells and so are useful adjunct in colon cancer therapy.

Sphingomyelin, one of the most abundant whey-derived sphingolipids, have potential to inhibit colon cancer. Sphingomyelins also regulate growth factor receptors, such as the transforming growth factor beta family (TGF-β).

TGF-βs are a multifunctional family of growth factors that regulate cell growth in normal and tumour cells by suppressing proliferation, inducing differentiation and apoptosis. TGF-β passes out unaffected through the stomach and maintains bioactivity in colon by withstanding enzymatic proteolysis.

Several scientific studies, based on experiments on animal experiments have shown the therapeutic effect of bovine lactoferrin (BLF) in treating distinct types of cancer (Gill & Cross, 2000 [41]), including colon cancer (Masuda et al., 2000 [42]). Besides this, the iron-binding capacity of bovine lactoferrin is responsible for its anticancer activity. The proposed mechanisms are that free iron act as a mutagenic promoter, by inducing oxidative damage to the nucleic acid structure; hence, when bovine lactoferrin binds iron in tissues, it reduces the risk of oxidant-induced carcinomas and colon adenocarcinomas. Studies are also available that pertain to such other cancers as lung, bladder, tongue and oesophagus, and which convey similar results (Masuda et al., 2000; Tanaka et al., 2000 [42,43]).

4.3. Immune-Enhancer

IgGs, Lp and Lf concentrated from whey participates in host immunity. IgGs binds with bacterial toxins and lowers the bacterial load in the large bowel. Dietary Lp and Lf plays a vital role in host immunity through their antibacterial action on pathogenic microorganisms. Thus, supplementing diets with whey-derived adjuncts shifts the balance of intestinal microflora and aids in immune enhancement. Calcium, by virtue of its probiotic activity and influence on macrophages, neutrophils and other white blood cell (WBC) subtypes are also classified as immune-modifying minerals.

Lf provides immune modulation as it is a natural component of the human immune system and is found in the body's mucous substances including saliva.

TGFs are abundant in milk (1–2 mg/L) and colostrum (20–40 mg/L) and serve a major role in GI integrity of newborn animals.

TGF-beta2 resides locally near the top of the intestinal villi, and inhibits the proliferation and signaling neutrophils as a part of innate host immunity.

Bovine lactoferrin upon stimulation of the immune system increases macrophage activity, causes induction of inflammatory cytokines, including IL-8, TNF-a and nitric oxide, stimulates proliferation of lymphocytes, and activates monocytes, natural killer (NK) cells and neutrophils (Gahr et al., 1991 [44]).

4.4. Prebiotic Properties

There are some whey-derived carbohydrate components that possess prebiotic activity. Lactose support LAB (such as *Bifidobacteria* & *Lactobacilli*). Stallic acids (type of oligosaccharides that are commonly found in whey), are typically attached to proteins that have been shown to possess prebiotic properties. Besides, three other non-carbohydrate prebiotics from whey are a protein called glycomacropeptide (GMP) that is derived from the partial enzymatic breakdown of kappa-casein during cheese production that supports the growth of bifidobacteria; second is lactoferrin (Lf), which supports the growth of bifidobacteria and lactobacilli and third component with prebiotic potential is mineral, calcium in form of calcium phosphate. It selectively stimulates the growth of intestinal lactobacilli and decreases the severity of *Salmonella* infections in rats (Kassem et al., 2015 [45]).

4.5. Anti-Inflammatory Properties

The role of whey proteins on hypertension is due to its effect on inflammation and Renin-Angiotensin System (RAS). The Angiotensin-Converting Enzyme (ACE) inhibitors possess anti-inflammatory properties (Sousa et al., 2012 [46]). In a study it was found that the consumption of whey proteins depleted the plasma levels of pro-inflammatory cytokines (IL-1 beta: 59% and IL-6: 29%) as compared to the consumption of same amount of casein. Thus, reduction of pro-inflammatory cytokines can also be associated with weight loss after consumption of whey proteins and its amino acids (Luhovyy et al., 2007 [47]).

4.6. Cardiovascular and Related Diseases

Cardiovascular disease (CVD) is associated with number of factors such as age, genetic constitution, obesity, sedentary lifestyle and alcohol intake and quality of dietary fat. Milk contains more than 12 different types of fat, including sphingolipids, free sterols, cholesterol and oleic acid. Several studies have shown that intake of milk and dairy products lower blood pressure and reduce the risk of hypertension (Groziak and Miller, 2000 [48]).

A study was undertaken to investigate the effect of fermented milk (*Lactobacillus casei* and *Streptococcus thermophiles*) supplement with an added whey protein concentrate on serum lipids and blood pressure on a group of 20 healthy adult males (Kawase et al., 2000 [49]). During the course of eight weeks, volunteers consumed 200 mL of fermented milk with whey protein concentrate or a placebo in the morning and evening. The placebo consisted of a non-fermented milk product without the addition of whey protein concentrate. After eight weeks, the fermented-milk group showed significantly higher High Density Lipoproteins (HDLs) and lower triglycerides and systolic blood pressure than did the placebo group.

In yet another study, the aim was to investigate the acute effects of dietary whey proteins on lipids, glucose and insulin, and resting energy expenditure in overweight and obese post-menopausal women, a population that is highly susceptible to cardiovascular disease. The findings suggested that a single dose of whey protein can decrease arterial exposure to smaller triglycerides-enriched lipoprotein particles compared to the glucose and casein meals in the postprandial period in overweight and obese, post-menopausal women (Pal et al., 2010 [50]).

4.7. GastroIntestinal Health

Whey proteins exert a therapeutic effect on the gastric mucosa. This effect is due to the presence of sulfhydryl group in aminoacid cysteine and its linkage with glutamic acid in the production of glutathione.

In a research by Rosaneli et al. [51], it was observed that when rats showed a 41% reduction in ulcerative lesions caused by ethanol ingestion when they were fed a whey protein concentrate, while a 73% reduction rate was observed following repeat doses of whey (McGregor and Poppitt, 2013 [52]).

Whey proteins are absorbed faster in body than casein. The lower absorption rate of casein in its native micellar form is due to the low pH conditions in the stomach that cause casein clotting and delays gastric emptying (Dangin et al., 2001 [53]). Therefore, plasma amino acids are more rapidly elevated following whey proteins consumption; whereas changes in plasma amino acids are lower and more sustained following micellar casein consumption. Thus, processing of whey proteins or casein fractions through hydrolysis can markedly influence absorption and subsequent plasma amino acid profiles.

4.8. Physical Strength and Performance for Athletes

Several studies have shown that whey proteins have tremendous nutraceutical potential for athletes. They are often referred to as 'gold standard' or 'fast protein' for their unique ability to provide nourishment to muscles. This is due to several factors such as their solubility, easily digestible, efficient absorption. They contain all the essential amino acids required in daily diet and have an ideal combination of amino acids to help to improve body composition and enhance athletic performance.

Whey protein is a rich source of branched chain amino acids that are important for athletes since they are metabolized directly into muscle tissue and repair and rebuild lean muscle tissue. Whey protein is an excellent source of leucine (an essential amino acid) that plays a key role in promoting muscle protein synthesis and muscle growth in athletes. It has been reported that whey protein isolate has approximately 50% more leucine than soy protein isolate. Whey protein increases the level of glutathione in the body and helps to maintain a healthy immune system (Gupta et al., 2012 [4]).

4.9. Obesity Control

Numerous studies have proved that whey proteins help in weight management and obesity control. The mechanism involved is through the influence of whey proteins on the hormones that controls appetite and hunger (Hall et al., 2003 [54]). Consumption of a high-protein diet has a greater satiety thereby reducing energy intake and subsequent adiposity. It has been found that whey proteins are even more effective than red meat in reducing weight gain and also increases insulin sensitivity of the body. The other proposed mode of action is that diet rich in α-lactalbumin enhances the lipid oxidation and rapidly delivers amino acids for use during exercise, thereby decreasing adiposity (Bouthegourd et al., 2002 [55]). Calcium is also known to influence energy metabolism as it regulates adipocyte lipid metabolism and triglyceride storage. This has been experimentally supported by Zemel 2002 [56] who demonstrated a greater effect of dairy versus non-dairy sources of calcium for improving body composition.

4.10. Protection against Human Immunodeficiency Virus (HIV)

Research has shown that individuals infected with HIV exhibit glutathione deficiency. Supplementation of whey elevates the levels of glutathione (GSH) and helps in increasing the immune defense mechanism of the host.

A research was undertaken by Micke et al. [57] on 30 subjects infected with HIV. They were randomized and received a daily dose of 45 g whey proteins from one of two sources Protectamin® or Immunocal. These two products are commercially available and possess different amino acid profiles. The oral supplementation was given for two weeks. After two weeks, the Protectamin-supplemented group demonstrated significantly elevated glutathione levels while the Immunocal group had statistically non-significant elevations. These results clearly showed that high levels of glutathione supplementation provide protection against HIV.

4.11. Diabetes

Numerous studies have proved that whey proteins help to control the blood glucose levels and provide additional benefits for weight management. This is a main concern for type-2 diabetics (Shankar and Bansal, 2013 [58]). Whey proteins are thus high biological, high-quality and value protein suitable for diabetic patients. Ingestion of whey proteins leads to more rapid secretion of insulin than micellar casein (McGregor and Poppitt, 2013 [52]).

4.12. Appetite Suppression

It has been mentioned earlier that whey based beverages regulates body weight and helps in controlling obesity. This is due to some amino acids liberated from whey proteins during *in vivo* digestion that also stimulate the release of hormones including insulin. Its secretion directly affects food intake regulation by modifying glycemic response and suppressing appetite and, consequently reduces body weight… Some other hormones are also involved in appetite suppression either directly in the hypothalamus, such as ghrelin, or indirectly through the vagal nerve, such as cholecystokinin (CCK) and Peptide YY (PYY) (Jakubowicz and Froy, 2013 [59]). Glyco-macropeptide is a powerful stimulator of CCK, which is an appetite-suppressing hormone that plays essential roles related to gastrointestinal function; including the regulation of food intake. Further research is required to study the effects of GMP and CCK as an appetite suppressant (Walzem, 1999 [60]).

4.13. Ageing

Numerous studies have shown that whey protein based drinks play an important role in delaying the ageing process. During ageing, there is a continuous but slow degeneration of skeletal muscle mass (sarcopenia). In elderly population the skeletal muscle mass is completely impaired and inhibited. Whey proteins are a rich source of essential amino acids that contributes to its anabolic properties. There is an increase in availability of postprandial plasma amino acids that stimulates synthesis of muscle proteins (Pennings et al., 2011 [61]). Whey proteins contain glutathione that is an important antioxidant involved in the maintenance of functional and structural integrity of muscular tissue undergoing oxidative damage during exercise and aging.

4.14. Wound Healing

Wound healing involves the growth of new skin through the use of proteins and their amino acids. Healing process is delayed when there are inadequate amounts of protein or diets high in poor-quality proteins, such as gelatin are present. Whey proteins comprises of good quality proteins and are therefore often recommended by physicians after any surgery or burn therapy (Gupta et al., 2012 [4]).

5. Therapeutic Applications of Fermented Whey

The therapeutic potential and functional properties of milk whey can be improved through fermentation (Yang and Silva, 1995 [62]). During fermentation by probiotic bacteria, the percentage of essential and digestible amino acids increases to a marked level, thus making these fermented dairy products ideal nutritional supplements for diarrhea and other conditions (Hitchins and Mc Donough, 1989 [63]). Various compounds including flavouring compounds and acids are produced due to the metabolic activity of these microorganisms. This also increases their palatability and consumer acceptability as compared to non-fermented products. The organic acids produced during the fermentation possess great health benefits. These organic acids comprise of short-chain fatty acids such as acetic, citric and lactic acid produced during the fermentation of protein and carbohydrate components. The organic acids possess antimicrobial properties particularly against *E. coli* in the intestine. The organic acids helps to lower the pH of the intestine that helps in the secretion of bile juices, absorption of nutrients and also reduce the concentration of pathogenic microglora in the gut. The SCFA is used in the colon to promote water and electrolyte absorption for diarrhoea treatment (Desjeux, 2000 [64]).

In a study in rats it was also reported that dietary calcium, preferably as calcium phosphate, also selectively favours the growth of intestinal lactobacilli and decreases the severity of *Salmonella* infections (Bovee-Oudenhoven et al., 1999 [65]).

In a study, Lactobacillus acidophilus subsp. *johnsonii* (La1) has been shown to effectively suppress the growth of *Helicobacter pylori* in vitro. The results were confirmed by conducting a hydrogen breath test. The whey-drink based on *L. acidophilus* (*johnsonii*) culture was given to some volunteers and a marked decrease in test values was observed (Michetti, 1999 [66]).

6. Whey-Based Probiotic Products

During recent years there has been great awareness about the consumption of dairy products amongst the consumers worldwide. Dairy products containing probiotic bacteria of selected strains preferably belonging from the group of *Lactobacillus* spp. and *Bifidobacterium* spp. are generally preferable. Fermented dairy products such as whey have a greater consumer appeal as they are nutritious, thirst-quenching, less acidic and low in calories. Since dairy products are highly perishable they are susceptible to microbial spoilage during storage. Nowadays, liquid whey is therefore concentrated by spray drying or by evaporation, ultrafiltration or reverse osmosis thereby increasing their shelf life.

Besides, these whey drinks are further fortified with the friendly probiotic bacteria for their health promoting properties, and provide unique texture and flavor to them. Some whey-based probiotic products available in the market are Yakult and Sofyl (manufactured by Yakult), Chamyto (by Nestlé), Activia, Actimel, Danito (by Danone), Vigor-Club (by Vigor) (Katz, 2001 [67]).

The most common method of producing whey drinks is to remove excess of whey during cheese manufacturing process. The liquid whey is then filtered, pasteurized and fermented with the desired probiotic strain. It has been commonly observed that sweet whey (prepared after coagulation with rennet) is tastier than acid whey. The whey obtained through this process is clearer and is thus not sedimented during long term storage. This is called deproteinised whey. Such whey drinks can also be easily carbonated similar to soft drinks due to their low viscosity. (Wilson and Temple, 2004 [68]).

In a study, when a whey based beverages fortified with probiotic culture of *Lactobacillus acidophilus* was administered to diarrheal children, positive results were reported (Goyal and Gandhi, 2008 [69]).

It is necessary that the whey based beverages should be fortified with only specific viable probiotic strain(s) in sufficient numbers and scientifically validated with appropriate labeling in order to ensure therapeutic results (Reid et al, 2006 [70]).

7. Conclusions

From the above discussion, it is imperative that milk whey possess tremendous therapeutic properties such as antimicrobial, anticancer, immune-enhancer, prebiotic property, anti-inflammatory, cardiovascular, gastro-intestinal health, physical strength, obesity control and weight-management, HIV, diabetes, appetite suppression, ageing, and wound healing. There are several whey-based probiotic products available in the market today that can serve as attractive health-promoting food supplements. Kefir, yogurts, frozen yogurts and desserts are all well-known examples.

Acknowledgments: The authors greatly acknowledge and thank respected Founder President Ashok K. Chauhan, and Atul Chauhan, Chancellor, Amity University UP, Noida, India, for their constant motivation, support and, research facilities.

Author Contributions: Charu Gupta has contributed towards data collection, drafting and compiling of the manuscript while Dhan Prakash has critically reviewed the manuscript for its accuracy and completeness.

Conflicts of Interest: The authors declare no conflict of interest.

Appendix A

Table A1. Nutrient content of cheese whey types (dry matter basis).

Nutrient Content	Sweet Whey (%)	Whey Permeate (%)
Total nitrogen (TN)	1.30	0.26
Non-protein nitrogen (NPN)	0.34	0.24
Calcium	0.058	0.055
Phosphorus	0.052	0.045
Net energy lactation (Mcal/lb)	0.90	0.85
Total digestible energy (Mcal/lb)	1.86	1.7

Table A2. Whey Characteristics.

Characteristics Chemical Composition	Sweet Whey	Whey Permeate
Specific gravity (kg/L)	1.025	1.030
pH	6.40	6.55
Titrable acidity	0.05	0.089
Water (%)	91.95	94.45
Dry matter (DM%)	8.05	5.55
- Solid not fat (SNF%)	7.55	5.55
- Fat (%)	0.50	0.00
Crude protein (CP%)	1.10	0.25
Soluble carbohydrates (%)	5.20	4.90
Total ash (%)	0.52	0.50

References

1. Yalcin, A.S. Emerging therapeutic potential of whey proteins and peptides. *Curr. Pharm. Des.* **2006**, *12*, 1637–1643. [CrossRef] [PubMed]
2. Sultan, S.; Huma, N.; Butt, M.S.; Aleem, M.; Abbas, M. Therapeutic potential of dairy bioactive peptides: A Contemporary Perspectives. *Crit. Rev. Food Sci. Nutr.* **2016**. [CrossRef] [PubMed]
3. Rognvaldardottir, N. *Icelandic Food and Cookery*; Hippocrene Books: New York, NY, USA, 2001.
4. Gupta, C.; Prakash, D.; Garg, A.P.; Gupta, S. Whey Proteins: A novel source of Bioceuticals. *Mid. East J. Sci. Res.* **2012**, *12*, 365–375.
5. Yolken, R.H.; Losonsky, G.A.; Vonderfecht, S.; Leister, F.; Wee, S.B. Antibody to human rotavirus in cow's milk. *N. Engl. J. Med.* **1985**, *312*, 605–610. [CrossRef] [PubMed]
6. Le Maux, S.; Giblin, L.; Croguennec, T.; Bouhallab, S.; Brodkorb, A. β-Lactoglobulin as a molecular carrier of linoleate: Characterization and effects on intestinal epithelial cells in vitro. *J. Agric. Food Chem.* **2012**, *60*, 9476–9483. [CrossRef] [PubMed]
7. Mehraban, M.H.; Yousefi, R.; Taheri-Kafrani, A.; Panahi, F.; Khalafi-Nezhad, A. Binding study of novel anti-diabetic pyrimidine fused heterocycles to β-lactoglobulin as a carrier protein. *Coll. Surf. B Biointerfaces* **2013**, *112*, 374–379. [CrossRef] [PubMed]
8. Brück, W.M.; Gibson, G.R.; Brück, T.B. The effect of proteolysis on the induction of cell death by monomeric alpha-lactalbumin. *Biochimie* **2014**, *97*, 138–143. [CrossRef] [PubMed]
9. Ganjam, L.S.; Thornton, W.H.; Marshall, R.T.; MacDonald, R.S. Antiproliferative effects of yoghurt fractions obtained by membrane dialysis on cultured mammalian intestinal cells. *J. Dairy Sci.* **1997**, *80*, 2325–2339. [CrossRef]
10. Markus, C.R.; Olivier, B.; de Haan, E.H. Whey protein rich in a-lactalbumin increases the ratio of plasma tryptophan to the sum of the other large neutral amino acids and improves cognitive performance in stress-vulnerable subjects. *Am. J. Clin. Nutr.* **2002**, *75*, 1051–1056. [PubMed]
11. Huang, B.X.; Kim, H.; Dass, C. Probing three-dimensional structure of bovine serum albumin by chemical cross-linking and mass spectrometry. *J. Am. Soc. Mass Spectrom.* **2004**, *15*, 1237–1247. [CrossRef] [PubMed]
12. Laursen, I.; Briand, P.; Lykkesfeldt, A.E. Serum albumin as a modulator of the human breast cancer cell line MCF-7. *Anticancer Res.* **1990**, *10*, 343–352. [PubMed]
13. Choi, J.K.; Ho, J.; Curry, S.; Qin, D.; Bittman, R.; Hamilton, J.A. Interactions of very long chain saturated fatty acids with serum albumin. *J. Lipid Res.* **2002**, *43*, 1000–1010. [CrossRef] [PubMed]
14. Tong, L.M.; Sasaki, S.; McClements, D.J.; Decker, E.A. Mechanisms of the antioxidant activity of a high molecular weight fraction of whey. *J. Agric. Food Chem.* **2000**, *48*, 1473–1478. [CrossRef] [PubMed]
15. Pierce, A.; Colavizza, D.; Benaissa, M.; Maes, P.; Tartar, A.; Montreuil, J.; Spik, G. Molecular cloning and sequence analysis of bovine lactotransferrin. *Eur. J. Biochem.* **1991**, *196*, 177–184. [CrossRef] [PubMed]
16. Levay, P.F.; Viljoen, M. Lactoferrin: A general review. *Haematologica* **1995**, *80*, 252–267. [PubMed]
17. El-Loly, M.M.; Mahfouz, M.B. Lactoferrin in Relation to Biological Functions and Applications: A Review. *Int. J. Dairy Sci.* **2011**, *6*, 79–111. [CrossRef]
18. El-Loly, M.M. Identification and Quantification of Whey Immunoglobulins by Reversed Phase Chromatography. *Int. J. Dairy Sci.* **2007**, *2*, 268–274.

19. Losso, J.N.; Dhar, J.; Kummer, A.; Li-Chan, E.; Nakai, S. Detection of antibody specificity of raw bovine and human milk to bacterial lipopolysaccharides using PCFIA. *Food Agric. Immunol.* **1993**, *5*, 231–239. [CrossRef]
20. Scammell, A.W. Production and uses of colostrum. *Aust. J. Dairy Technol.* **2001**, *56*, 74–82.
21. Garry, F.B.; Adams, R.; Cattell, M.B.; Dinsmore, R.P. Comparison of passive immunoglobulin transfer to dairy calves fed colostrum or commercially available colostral supplement products. *JAVMA* **1996**, *208*, 107–110. [PubMed]
22. Hilpert, H.; Brussow, H.; Mietens, C.; Sidoti, J.; Lerner, L.; Werchau, H. Use of bovine milk concentrate containing antibody to rotavirus to treat rotavirus gastroenteritis in infants. *J. Infect. Dis.* **1987**, *156*, 158–166. [CrossRef] [PubMed]
23. Korhonen, H.; Marnila, P.; Gill, H.S. Milk immunoglobulins and complement factors. *Br. J. Nutr.* **2000**, *84*, S75–S80. [CrossRef] [PubMed]
24. Lilius, E.M.; Marnila, P. The role of colostral antibodies in prevention of microbial infections. *Curr. Opin. Infect. Dis.* **2001**, *14*, 295–300. [CrossRef] [PubMed]
25. De Wit, J.N.; van Hooydonk, A.C.M. Structure, functions and applications of lactoperoxidase in natural antimicrobial systems. *Neth. Milk Dairy J.* **1996**, *50*, 227–244.
26. Reiter, B.; Perraudin, J.P. Lactoperoxidase, Biological Functions. In *Peroxidases in Chemistry and Biology*; Everse, J., Everse, K.E., Grisham, M.B., Eds.; CRC Press: Boca Raton, FL, USA, 1991; pp. 144–180.
27. Shin, K.; Wakabayashi, H.; Yamauchi, K.; Teraguchi, S.; Tamura, Y.; Kurokawa, M.; Shiraki, K. Effects of orally administered bovine lactoferrin and lactoperoxidase on influenza virus infection in mice. *J. Med. Microbiol.* **2005**, *54*, 717–723. [CrossRef] [PubMed]
28. Marshall, K. Therapeutic applications of whey protein. *Altern. Med. Rev.* **2004**, *9*, 136–156. [PubMed]
29. Gustafson, D.R.; McMahon, D.J.; Morrey, J.; Nan, R. Appetite is not influenced by a unique milk peptide: Caseinomacropeptide (CMP). *Appetite* **2001**, *36*, 157–163. [CrossRef] [PubMed]
30. Fox, P.F.; Kelly, A.L. Indigenous enzymes in milk: Overview and historical aspects-Part 2. *Int. Dairy J.* **2006**, *16*, 517–532. [CrossRef]
31. Lesnierowsk, G. New manners of physical-chemical modification of lysozyme. *Nauka Przyr. Technol.* **2009**, *3*, 1–18.
32. Benkerroum, N. Antimicrobial activity of lysozyme with special relevance to milk. *Afr. J. Biotechnol.* **2008**, *7*, 4856–4867.
33. Krissansen, G.W. Emerging health properties of whey proteins and their clinical implications. *J. Am. Coll. Nutr.* **2007**, *26*, 713–723. [CrossRef]
34. Troost, F.J.; Steijns, J.; Saris, W.H.; Brummer, R.J. Gastric digestion of bovine lactoferrin in vivo in adults. *J. Nutr.* **2001**, *131*, 2101–2104. [PubMed]
35. Leães, F.L.; Vanin, N.G.; Sant'Anna, V.; Brandelli, A. Use of Byproducts of Food Industry for Production of Antimicrobial Activity by Bacillus sp. P11. *Food Bioprocess Technol.* **2010**, *4*, 822–828. [CrossRef]
36. Chatterton, D.E.W.; Smithers, G.; Roupas, P.; Brodkrob, A. Bioactivity of β-lactoglobulin and α-lactalbumin-technological implications for processing. *Int. Dairy J.* **2006**, *16*, 1229–1240. [CrossRef]
37. Takakura, N.; Wakabayashi, H.; Ishibashi, H.; Teraguchi, S.; Tamura, Y.; Yamaguchi, H.; Abe, S. Oral lactoferrin treatment of experimental oral candidiasis in mice. *Antimicrob. Agents Chemother.* **2003**, *47*, 2619–2623. [CrossRef] [PubMed]
38. Floris, R.; Recio, I.; Berkhout, B.; Visser, S. Antibacterial and antiviral effects of milk proteins and derivatives thereof. *Curr. Pharm. Des.* **2003**, *9*, 1257–1275. [CrossRef]
39. Wolber, F.M.; Broomfield, A.M.; Fray, L.; Cross, M.L.; Dey, D. Supplemental dietary whey protein concentrate reduces rotavirus-induced disease symptoms in suckling mice. *J. Nutr.* **2005**, *135*, 1470–1474. [PubMed]
40. Solak, B.B.; Akin, N. Health benefits of whey protein: A review. *J. Food Sci. Eng.* **2012**, *2*, 129–137.
41. Gill, H.S.; Cross, M.L. Anticancer properties of bovine milk. *Br. J. Nutr.* **2000**, *84*, S161–S166. [CrossRef] [PubMed]
42. Masuda, C.; Wanibushi, H.; Sekine, K.; Yano, Y.; Otani, S.; Kishimoto, T.; Fukushima, S. Chemo-preventive effect of bovine lactoferrin on N-butyl-N-(4-hydroxybutyl)-nitrosamine-induced bladder carcinogenesis. *Jpn. J. Cancer Res.* **2000**, *91*, 582–588. [CrossRef] [PubMed]
43. Tanaka, T.; Kawabata, K.; Kohno, H.; Honjo, S.; Murakami, M.; Ota, T.; Tsuda, H. Chemo-preventive effect of bovine lactoferrin on 4-nitroquinoline 1-oxide induced tongue carcinogenesis in male F344 rats. *Jpn. J. Cancer Res.* **2000**, *91*, 25–33. [CrossRef] [PubMed]

44. Gahr, M.; Speer, C.P.; Damerau, B.; Sawatzki, G. Influence of lactoferrin on the function of human polymorphonuclear leukocytes and monocytes. *J. Leucoc. Biol.* **1991**, *49*, 427–433.

45. Kassem, J.M. Future Challenges of Whey Proteins. *Int. J. Dairy Sci.* **2015**, *10*, 139–159. [CrossRef]

46. Sousa, G.T.D.; Lira, F.S.; Rosa, J.C.; de Oliveira, E.P.; Oyama, L.M.; Santos, R.V.; Pimente, G.D. Dietary whey protein lessens several risk factors for metabolic diseases: A review. *Lipids Health Dis.* **2012**, *11*, 67. [CrossRef] [PubMed]

47. Luhovyy, B.L.; Akhavan, T.; Anderson, G.H. Whey proteins in the regulation of food intake and satiety. *J. Am. Coll. Nutr.* **2007**, *26*, 704S–712S. [CrossRef] [PubMed]

48. Groziak, S.M.; Miller, G.D. Natural bioactive substances in milk and colostrum: effects on the arterial blood pressure system. *Br. J. Nutr.* **2000**, *84*, S119–S125. [CrossRef] [PubMed]

49. Kawase, M.; Hashimoto, H.; Hosoda, M.; Morita, H.; Hosono, A. Effect of administration of fermented milk containing whey protein concentrate to rats and healthy men on serum lipids and blood pressure. *J. Dairy Sci.* **2000**, *83*, 255–263. [CrossRef]

50. Pal, S.; Ellis, V.; Ho, S. Acute effects of whey protein isolate on cardiovascular risk factors in overweight, post-menopausal women. *Atherosclerosis* **2010**, *212*, 339–344. [CrossRef] [PubMed]

51. Rosaneli, C.F.; Bighetti, A.E.; Antonio, M.A.; Carvalho, J.E.; Sgarbieri, V.C. Efficacy of a Whey protein concentrate on the inhibition of stomach ulcerative Lesions caused by ethanol ingestion. *J. Med. Food* **2002**, *5*, 221–228. [CrossRef] [PubMed]

52. McGregor, R.A.; Poppitt, S.D. Milk protein for improved metabolic health: A review of the evidence. *Nutr. Metab.* **2013**, *10*, 46. [CrossRef] [PubMed]

53. Dangin, M.; Boirie, Y.; Garcia-Rodenas, C.; Gachon, P.; Fauquant, J.; Callier, P.; Ballèvre, O.; Beaufrère, B. The digestion rate of protein is an independent regulating factor of postprandial protein retention. *Am. J. Physiol. Endocrinol. Metab.* **2001**, *280*, E340–E348. [PubMed]

54. Hall, W.L.; Millward, D.J.; Long, S.J.; Morgan, L.M. Casein and whey exert different effects on plasma amino acid profiles, gastrointestinal hormone secretion and appetite. *Br. J. Nutr.* **2003**, *89*, 239–248. [CrossRef] [PubMed]

55. Bouthegourd, J.C.J.; Roseau, S.M.; Makarios-Lahham, L.; Leruyet, P.M.; Tome, D.G.; Even, P.C. A preexercise α-lactalbumin-enriched whey protein meal preserves lipid oxidation and decreases adiposity in rats. *Am. J. Physiol. Endocrinol. Metab.* **2002**, *283*, E565–E572. [CrossRef] [PubMed]

56. Zemel, M.B. Mechanisms of dairy modulation of adiposity. *J. Nutr.* **2002**, *133*, 252–256.

57. Micke, P.; Beeh, K.M.; Buhl, R. Effects of long-term supplementation with whey proteins on plasma glutathione levels of HIV-infected patients. *Eur. J. Nutr.* **2002**, *41*, 12–18. [CrossRef] [PubMed]

58. Shankar, J.R.; Bansal, G.K. A study on health benefits of whey proteins. *Int. J. Adv. Biotechnol. Res.* **2013**, *4*, 15–19.

59. Jakubowicz, D.; Froy, O. Biochemical and metabolic mechanisms by which dietary whey protein may combat obesity and Type 2 diabetes. *J. Nutr. Biochem.* **2013**, *24*, 1–5. [CrossRef] [PubMed]

60. Walzem, R.L. *Health Enhancing Properties of Whey Proteins and Whey Fractions*; USA Dairy Export Council: Arlington, VA, USA, 1999.

61. Pennings, B.; Boirie, Y.; Senden, J.M.G.; Gijsen, A.P.; Kuipers, H.; van Loon, L.J.C. Whey protein stimulates postprandial muscle protein accretion more effectively than do casein and casein hydrolysate in older men. *Am. J. Clin. Nutr.* **2011**, *93*, 997–1005. [CrossRef] [PubMed]

62. Yang, S.T.; Silva, E.M. Novel products and new technologies for use of a familiar carbohydrate, milk lactose. *J. Dairy Sci.* **1995**, *78*, 2541–2562. [CrossRef]

63. Hitchins, A.D.; Mc Donough, F.E. Prophylactic and therapeutic aspects of fermented milk. *Am. J. Clin. Nutr.* **1989**, *49*, 675–684. [PubMed]

64. Desjeux, J.F. Can mal-absorbed carbohydrates be useful in the treatment of acute diarrhoea? *J. Pediatr. Gastroenterol. Nutr.* **2000**, *31*, 503–507. [CrossRef]

65. Bovee-Oudenhoven, I.M.; Wissink, M.L.; Wouters, J.T.; Van der Meer, R. Dietary calcium phosphate stimulates intestinal lactobacilli and decreases the severity of *Salmonella* infections. *J. Nutr.* **1999**, *129*, 607–612. [PubMed]

66. Michetti, P.; Dorta, G.; Wiesel, P.H.; Brassart, D.; Verdu, E.; Herranz, M.; Felley, C.; Porta, N.; Rouvet, M.; Blum, A.L.; et al. Effect of whey based culture supernatant of *Lactobacillus acidophilus* (*johnsonii*) LA1 on the *Helicobacter pylori* infection in human. *Digestion* **1999**, *60*, 203–209. [CrossRef] [PubMed]

67. Katz, F. Active cultures add function to yogurt and other foods. *Food Technol.* **2001**, *55*, 46–49.

68. Wilson, T.; Temple, N.J. *Beverages in Nutrition and Health*; Humana Press: New York, NY, USA, 2004; p. 427, ISBN 1-59259-415-8.
69. Goyal, N.; Gandhi, D.N. Whey, a Carrier of Probiotics against Diarrhoea. Available online: http://www.dairyscience.info/probiotics/110-whey-probiotics.html?showall=1 (accessed on 17 May 2011).
70. Reid, G.; Kim, S.O.; Kohler, G.A. Selecting, testing and understanding probiotic microorganisms. *FEMS Immunol. Med. Microbiol.* **2006**, *46*, 149–157. [CrossRef] [PubMed]

beverages

MDPI

Article

Dietary Milk Sphingomyelin Reduces Systemic Inflammation in Diet-Induced Obese Mice and Inhibits LPS Activity in Macrophages

Gregory H. Norris [†], Caitlin M. Porter [†], Christina Jiang and Christopher N. Blesso *

Department of Nutritional Sciences, University of Connecticut, Storrs, CT 06269, USA;
gregory.norris@uconn.edu (G.H.N.); caitlin.porter@uconn.edu (C.M.P.); christina.jiang@uconn.edu (C.J.)
* Correspondence: christopher.blesso@uconn.edu; Tel.: +1-860-486-9049
† Both investigators contributed equally to this work.

Academic Editor: Alessandra Durazzo
Received: 31 May 2017; Accepted: 14 July 2017; Published: 21 July 2017

Abstract: High-fat diets (HFD) increase lipopolysaccharide (LPS) activity in the blood and may contribute to systemic inflammation with obesity. We hypothesized that dietary milk sphingomyelin (SM), which reduces lipid absorption and colitis in mice, would reduce inflammation and be mediated through effects on gut health and LPS activity. C57BL/6J mice were fed high-fat, high-cholesterol diets (HFD, $n = 14$) or the same diets with milk SM (HFD-MSM, 0.1% by weight, $n = 14$) for 10 weeks. HFD-MSM significantly reduced serum inflammatory markers and tended to lower serum LPS ($p = 0.08$) compared to HFD. Gene expression related to gut barrier function and macrophage inflammation were largely unchanged in colon and mesenteric adipose tissues. Cecal gut microbiota composition showed greater abundance of *Acetatifactor* genus in mice fed milk SM, but minimal changes in other taxa. Milk SM significantly attenuated the effect of LPS on pro-inflammatory gene expression in RAW264.7 macrophages. Milk SM lost its effects when hydrolysis was blocked, while long-chain ceramides and sphingosine, but not dihydroceramides, were anti-inflammatory. Our data suggest that dietary milk SM may be effective in reducing systemic inflammation through inhibition of LPS activity and that hydrolytic products of milk SM are important for these effects.

Keywords: sphingomyelin; sphingolipids; ceramide; sphingosine; milk; dairy; obesity; inflammation; gut; macrophage

1. Introduction

Chronic low-grade inflammation is involved in the pathogenesis of cardiovascular disease, type 2 diabetes, and non-alcoholic fatty liver disease [1]. Metabolically related chronic low-grade inflammation may be exacerbated by circulating lipopolysaccharide (LPS), also called "endotoxin", a pro-inflammatory molecule found in the outer membrane of Gram-negative bacteria [2]. Recognition of bacterial LPS by host cells can trigger a pro-inflammatory signaling cascade mediated by Toll-like receptor-4 (TLR4), a pattern recognition receptor [3]. The human gastrointestinal tract has a "gut barrier" to limit the passage of microbes between host intestinal cells (e.g., tight junctions between enterocytes) and into the bloodstream [4]. These features of the gut barrier reduce permeability to LPS, which may otherwise trigger host inflammation and disease [5]. However, LPS translocation from the gut to circulation is enhanced by the absorption of dietary lipids [6]. High-fat diets (HFD) have been shown to promote inflammation of the distal intestine, causing an impairment of the protective gut barrier, and translocation of LPS into host circulation [7]. Diets high in fat can also alter the gut microbiota [8], resulting in diminished amounts of beneficial bacteria, such as some bifidobacteria, which would promote gut barrier permeability [7]. Additionally, diets rich in triglyceride and cholesterol increase chylomicron production during digestion and absorption. It has been shown that LPS can be incorporated into chylomicrons and transported into the bloodstream of mice [6]. Overall, these effects of HFD result in the paracellular and transcellular translocation of LPS from the gut into the circulation. The presence of LPS in the circulation activates the inflammatory response of the immune system, mainly through the activation of TLR4 [3]. Lipopolysaccharide engages the TLR4/MD-2 receptor complex via LPS-binding protein (LBP) and CD14. Toll-like receptor 4 signaling then activates the nuclear factor-kappa B (NF-κb) transcription factor resulting in the production of pro-inflammatory cytokines, such as tumor necrosis factor alpha (TNF-α) [3]. Macrophages in the colon have been shown to be the first pro-inflammatory immune responders to an HFD [9]; thus, inhibiting LPS translocation and the subsequent inflammatory responses of macrophages shows great potential in reducing the detrimental effects of a diet rich in fat and cholesterol.

Dietary phospholipids, including sphingolipids, show potential in mitigating chronic disease through effects on lipid absorption and inflammation [10,11]. Dietary sphingolipids, which include sphingomyelin (SM), ceramides, and sphingosine, are mainly found in milk, eggs, and soybeans [12]. It is estimated that the average American consumes 0.3 to 0.4 grams of sphingolipids per day [12]. Sphingomyelin is considered a zoochemical, being present in animal cell membranes but absent from plants [13]. Sphingomyelin found in milk is an important component of milk fat globule membranes [14]. Dietary SM and other sphingolipids have been studied for their effects on dyslipidemia because they interfere with the absorption of dietary fat and cholesterol [15–21]. In addition to effects on lipid absorption, dietary SM may have further bioactive effects by reducing inflammation and LPS activity, potentially influencing chronic disease. Dietary milk SM has been shown to reduce dextran sulfate sodium (DSS)-induced colitis in mice, suggesting anti-inflammatory effects in the gut [22]. However, the effect of dietary SM on colon inflammation is controversial, as exacerbation of colitis in mice has been observed with feeding egg-derived SM [23,24]. Phospholipids and sphingolipids appear to impact LPS activity, as they have been shown to dampen LPS-induced inflammation [25,26]. The presence of LPS in circulation also results in the liver upregulating sphingolipid biosynthesis via the sphingolipid rate-limiting biosynthetic enzyme, serine palmitoyltransferase (SPT), possibly as a compensatory mechanism [27]. We have previously shown that feeding 0.25% (w/w) milk SM reduced serum LPS and altered the gut microbiota in HFD-fed mice after four weeks [28]. We also recently reported that feeding dietary SM (0.1% w/w) attenuated hepatic steatosis and adipose tissue inflammation in diet-induced obese mice [29]. The current study aimed to assess the effects of dietary milk SM on systemic and gut inflammation using a HFD-induced obese mouse model. We also sought to elucidate whether milk SM and its hydrolytic products (ceramides, dihydroceramides, and sphingosine) directly affect the inflammatory response of macrophages stimulated by LPS.

2. Animals and Diets

Male C57BL/6J mice were purchased from the Jackson Laboratory (Bar Harbor, ME, USA) at six weeks of age. Mice were housed in a temperature-controlled room and maintained in a 12 h light/12 h dark cycle within the University of Connecticut-Storrs vivarium. The Animal Care and Use Committee of the University of Connecticut-Storrs approved all procedures used in the current study. Mice were acclimated to the facility for two weeks before being placed on either a lard-based high-fat, high-cholesterol diet (HFD; 60% kcal from fat, 0.15% cholesterol added by weight; $n = 14$) or HFD supplemented with 0.1% of milk SM added by weight (HFD-MSM; $n = 14$). Detailed diet compositions are presented in Table 1. Experimental diets were prepared using purified ingredients commercially available from Dyets, Inc. (Bethlehem, PA, USA). Milk SM (bovine; >99% purity) was obtained from Avanti Polar Lipids, Inc. (Alabaster, AL, USA) and then substituted for an equal weight of lard in the treatment group. Accounting for weight gained throughout the study, HFD-MSM provided the equivalent of consuming approximately 405–670 mg milk SM/day in a 70 kg human based on body surface normalization [30].

Table 1. Diet Composition.

Diet Component (g/kg of Diet)	High-Fat Diet	0.1% Milk Sphingomyelin High-Fat Diet
Casein	265	265
L-Cystine	4	4
Corn Starch	0	0
Maltodextrin	0	0
Sucrose	253.5	253.5
Lard	310	309
Soybean Oil	30	30
Cellulose	64	64
Mineral Mix, AIN-93G-MX (94046)	48	48
Vitamin Mix, AIN-93-VX (94047)	21	21
Choline Bitartrate	3	3
Cholesterol	1.5	1.5
Milk Sphingomyelin	0	1

Mice consumed the diets *ad libitum* and fresh food was provided twice per week. Body weight was assessed weekly, while food intake was calculated at each feeding. After 10 weeks, mice were fasted for 6–8 h followed by euthanasia and blood collection via cardiac puncture. Blood was allowed to clot at room temperature for 30 min before serum isolation by centrifugation ($10,000 \times g$ for 10 min at 4 °C) and then stored at −80 °C. Special precautions were taken to minimize endotoxin contamination of the serum samples. Mesenteric adipose, small intestine, and colon tissues were isolated and snap-frozen in liquid nitrogen before storage in −80 °C.

2.1. Serum Biochemical Analysis

Serum IL-6, TNF-α, IFNγ, and MIP-1β were measured by Luminex/xMAP magnetic bead-based multiplexing assay using MAGPIX instrumentation from EMD Millipore (Billerica, MA, USA). Serum LPS was measured using a chromogenic limulus amebocyte lysate (LAL) assay (QCL-1000) obtained from Lonza (Basel, Switzerland).

2.2. Gut Microbiota Analysis

Cecal fecal samples were collected from mice and submitted to the University of Connecticut-Storrs Microbial Analysis, Resources, and Services (MARS) facility for microbiota characterization utilizing 16S V4 analysis. DNA was extracted from 0.25 g of fecal sample using the MoBio PowerMag Soil 96 well kit (MoBio Laboratories, Inc., Carlsbad, CA, USA) according to the manufacturer's protocol for the Eppendorf epMotion liquid handling robot. DNA extracts were quantified using the Quant-iT PicoGreen kit (ThermoFisher Scientific, Waltham, MA, USA). Partial bacterial 16S rRNA (V4) and fungal ITS2 genes were amplified using 30 ng extracted DNA as template. The V4 region was amplified using 515F and 806R with Illumina adapters and dual indices (8 basepair golay on 3′ [31], and 8 basepair on the 5′ [32]). Samples were amplified in triplicate using Accuprime PFX PCR master mix (ThermoFisher Scientific, Waltham, MA, USA) with the addition of 10 μg BSA (New England BioLabs, Ipswich, MA, USA). The PCR reaction was incubated at 95 °C for 2 min., then 30 cycles of 15 s at 95.0 °C, 1 min. at 55 °C, and 1 min. at 68 °C, followed by final extension at 68 °C for 5 min. PCR products were pooled for quantification and visualization using the QIAxcel DNA Fast Analysis (Qiagen, Hilden, Germany). PCR products were normalized based on the concentration of DNA from 250–400 bp then pooled using the QIAgility liquid handling robot. The pooled PCR products were cleaned using Mag-Bind RXNPure Plus (Omega Bio-Tek, Norcross, GA, USA) according to the manufacturer's protocol. The cleaned pool was sequenced on the MiSeq using v2 2 x 250 base pair kit (Illumina, Inc, San Diego, CA, USA).

Sequences were demultiplexed using onboard bcl2fastq. Demultiplexed sequences were processed in Mothur v. 1.39.4 following the MiSeq SOP [32]. Exact commands can be found at [33]. Merged sequences that had any ambiguities or did not meet length expectations were removed. Sequences were aligned to the Silva nr_v119 alignment [34]. Identification of operational taxonomic units (OTUs) was done using the RDP Bayesian classifier [35] against the Silva nr_v119 taxonomy database.

2.3. Cell Culture

Murine RAW264.7 macrophages were obtained from ATCC (Manassas, VA, USA) and cultured in a humidified incubator at 37 °C with 5% CO_2. Cells were maintained in Dulbecco's modified Eagle's medium (4 g/L glucose) containing sodium pyruvate, 10% fetal bovine serum (Hyclone, Logan, UT, USA), 2 mM L-glutamine, 100 U/mL penicillin, 100 μg/mL streptomycin antibiotic (ThermoFisher Scientific, Waltham, MA, USA), and 100 μg/mL Normocin (Invitrogen, Carlsbad, CA, USA). Cell counts and viability were routinely measured using Trypan blue and a TC-20 automated cell counter (Bio-Rad, Hercules, CA, USA).

2.4. Effects of Sphingolipids on LPS Stimulation of RAW264.7 Macrophages

Milk sphingomyelin (bovine), sphingosine (d18:1), C16-ceramide (d18:1/16:0), C24-ceramide (d18:1/24:0), C16-dihydroceramide (d18:0/16:0), and C24-dihydroceramide (d18:0/24:0) were obtained from Avanti Polar Lipids at >99% purity. Milk SM and sphingosine were dissolved in 100% ethanol (EtOH), while ceramides were dissolved in EtOH/dodecane (98.8/0.2, v/v). RAW264.7 cells were seeded in 24-well plates and allowed to adhere overnight prior to experimentation. To examine possible anti-inflammatory effects of milk SM, RAW264.7 macrophages were pre-incubated with milk SM (0.8–8 μg/mL) or EtOH vehicle control for 1 h, followed by a 4 h co-incubation in the presence or absence of 1 ng/mL LPS (*E. coli* 0111:B4) (Sigma-Aldrich, St. Louis, MO, USA). Since milk SM comprised of a natural mixture of SM species, concentrations are reported in 0.8–8 μg/mL, which is comparable to 1–10 μM of a pure SM species. To determine if effects were dependent on SM hydrolysis, cells were incubated for 2 h with 100 μM imipramine (Sigma-Aldrich, St. Louis, MO, USA) prior to 1 h incubation with milk SM or control ± 4 h co-incubation with LPS. Imipramine has been shown to reduce acid sphingomyelinase (SMase) activity to ~20% of basal levels after 2 h of incubation [36]. To test the effects of various ceramides and sphingosine, RAW264.7 macrophages were incubated for 4 h with ceramides (10 μM), dihydroceramides (10 μM), sphingosine (1–10 μM), or vehicle control ± LPS (1 ng/mL). Ethanol was used as a vehicle control for milk SM and sphingosine treatments, while EtOH/dodecane (98.8/0.2, v/v) was used as a control for ceramide treatments. To test for cytotoxicity, Cell Counting Kit-8 (CCK-8) (Sigma-Aldrich, St. Louis, MO, USA) was used after cells were treated with the different sphingolipids. Sodium dodecyl sulfate (SDS, 50 μM) was used as a cytotoxic positive control. Caspase 3 activity was also determined as an indicator of apoptosis using a colorimetric protease assay kit (ThermoFisher Scientific, Waltham, MA, USA) according to the manufacturer's instructions. Cisplatin (ThermoFisher Scientific, Waltham, MA, USA) at 50 μM concentration was used on cells as a positive control to induce apoptosis.

2.5. RNA isolation, cDNA Synthesis, and qRT-PCR

Total RNA from small intestine, colon, mesenteric adipose tissue, and RAW264.7 macrophages was isolated using TRIzol reagent (Life Technologies, Carlsbad, CA, USA). RNA was treated with DNase I before reverse transcription using iScript cDNA synthesis kit (Bio-Rad). Real-time qRT-PCR using the iTaq Universal SYBR Green Supermix was performed on a CFX96 real-time-PCR detection system (Bio-Rad). For small intestine, colon, and mesenteric adipose tissues, the geometric mean of the reference genes, glyceraldehyde 3-phosphate dehydrogenase (Gapdh) and beta actin was used to standardize mRNA expression using the $2^{-\Delta\Delta Ct}$ method. The macrophage mRNA was standardized to Gapdh expression. All primer sequences used are listed in Table S1.

2.6. Statistical Analysis

Data were analyzed using GraphPad Prism 6. Means were compared using Student's *t* test or one-way analysis of variance with Holm–Sidak post hoc analysis where indicated. For gut microbiota

Beverages **2017**, *3*, 37

analysis, alpha and beta diversity statistics were calculated by taking the average of 1000 random subsampling to 10,000 reads per sample in Mothur. Non-metric multidimensional scaling (NMS) and Permanova were run in R 3.3.2. A subsampled species matrix was used for indicator species analysis [37]. Figures were drawn in R 3.3.2 using ggplot2 2.2.1 and RColorBrewer 1.1-2. Data are reported as mean ± SEM. Significance is reported as $p < 0.05$.

3. Results

3.1. Dietary Milk SM Reduces Systemic Inflammation and Tends to Lower Circulating LPS

All mice fed the lard-based HFD were obese and insulin-resistant after 10 weeks, while supplementation of 0.1% (w/w) milk SM did not affect food intake, body or tissue weight gain compared to HFD control [29]. However, feeding milk SM strongly decreased serum inflammatory cytokines/chemokines compared to HFD control (Figure 1A). Milk SM tended to reduce LPS compared to HFD control (-36%), but it was not significantly different ($p = 0.08$) (Figure 1B). Milk SM also tended to increase Niemann-Pick C1-Like 1 (NPC1L1) mRNA expression ($p = 0.07$) in the small intestine (Figure 2A), which is induced by cellular cholesterol depletion [38]. No other significant changes were observed in the small intestine for gene expression related to lipid absorption. Interestingly, C-C motif chemokine ligand 2 (CCL2) mRNA expression was significantly increased in the colon by milk SM (Figure 2B). However, this did not appear to alter gut health, as mRNA expression of genes related to macrophage infiltration/inflammation (F4/80, Cd68, Cd11c, Tnf) and gut barrier function (Tjp1, Alpi, Ocln) was mostly unaffected by milk SM (Figure 2B). Furthermore, mesenteric adipose tissue mRNA expression related to inflammation was also unchanged, while there was a trend in GLUT4 mRNA expression to be increased by milk SM ($p = 0.09$) (Figure 2C).

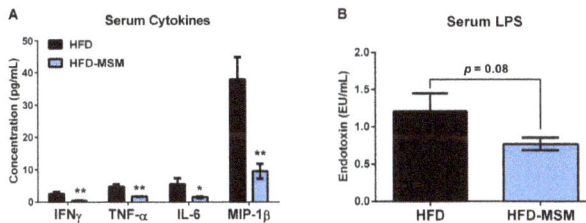

Figure 1. Dietary milk sphingomyelin (SM) reduces serum inflammation markers in diet-induced obese mice. Male C57BL/6J mice were fed high-fat diet (60% kcal fat, 31% lard, 0.15% added cholesterol) (high-fat diet (HFD) or HFD supplemented with 0.1% (w/w) milk SM (HFD-MSM) for 10 weeks. Serum cytokines/chemokines (**A**) and endotoxin concentrations (**B**) were measured by magnetic bead-based assay and chromogenic limulus amebocyte lysate (LAL) assay, respectively. Mean ± SEM, $n = 13–14$ per group. * $p < 0.05$, ** $p < 0.01$ compared to HFD.

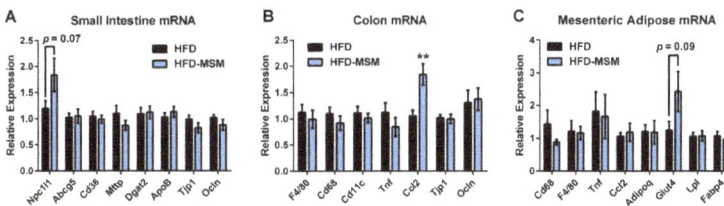

Figure 2. Intestinal and mesenteric adipose tissue mRNA expression of diet-induced obese mice. Small intestine mRNA (**A**), colon mRNA (**B**), and mesenteric adipose tissue mRNA (**C**) was determined using real-time qRT-PCR and standardized to the geometric mean of Gapdh and β-actin reference genes using the $2^{(-\Delta\Delta Ct)}$ method. Mean ± SEM, $n = 13–14$ per group. ** $p < 0.01$ compared to HFD.

3.2. Gut Microbiota Composition is Mostly Unaffected by 0.1% (w/w) Dietary Milk SM

To determine if alterations in gut microbiota could explain differences in systemic inflammation in the present study, cecal feces microbiota composition was examined by 16S rRNA sequence analysis and results are presented in Figure 3. *Verrucomicrobia* was the major phylum in cecal feces, but there were no significant differences between groups in its relative abundance or that of its major genus, *Akkermansia* (Figure 3B). Additionally, no significant differences were observed in the relative abundance of *Firmicutes* or *Bacteroidetes* phyla between groups (Figure 3B) or when phyla were grouped as Gram-negative bacteria (HFD: 76.9% ± 1.5% vs. HFD-MSM: 73.5% ± 2.8%, $p = 0.31$). However, relative abundance of *Acetatifactor*, a genus of the *Firmicutes* phylum, was significantly higher in the group fed milk SM compared to HFD control (Figure 3B). Alpha-diversity analysis (diversity within a sample) determined by inverse Simpson index was not significantly different between groups (HFD: 2.19 ± 0.16 vs. HFD-MSM: 2.46 ± 0.32, $p = 0.45$). Furthermore, beta-diversity analysis (between sample diversity) by Bray–Curtis revealed no significant clustering of samples according to diet (Figure 3C). Therefore, in the context of diet-induced obesity, gut microbiota composition was generally unaffected by supplementation of 0.1% (w/w) milk SM in the diet.

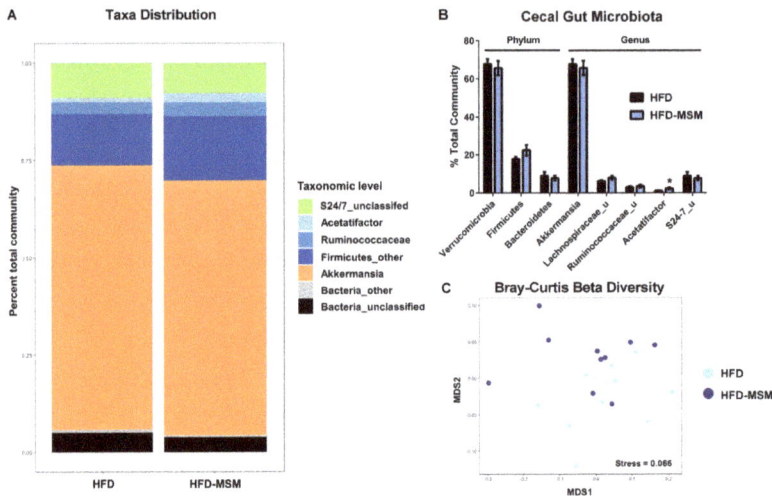

Figure 3. Gut microbiota composition of diet-induced obese mice. Cecal fecal samples were collected and microbiota composition was assessed by 16S rRNA sequencing as described in Materials and Methods. Mean phylogenetic abundance (% of total sequences) of mice (**A**) and taxa comparisons (**B**). Beta diversity shown by Bray–Curtis-based non-metric multidimensional scaling plot to visualize between sample diversity (**C**). Mean ± SEM, $n = 10$ per group. * $p < 0.05$ compared to HFD. Abbreviations: Lachnospiraceae_u, Lachnospiraceae_unclassified; Ruminococcaceae_u, Ruminococcaceae_unclassified; S24-7_u, S24-7 unclassified.

3.3. Milk SM Inhibits LPS Stimulation of Macrophages

Since dietary milk SM may directly influence inflammatory processes beyond inhibiting lipid absorption, we examined its effects on pro-inflammatory gene expression in RAW264.7 macrophages. Increases in TNF-α and CCL2 mRNA with 4 h LPS stimulation were both significantly reduced with milk SM (0.8 and 8 μg/mL) (Figure 4A,B). Milk SM in the absence of LPS did not influence pro-inflammatory gene expression. Furthermore, cell viability was not affected by milk SM at concentrations shown to affect pro-inflammatory gene expression (Figure 4C). With the addition of imipramine, LPS-stimulated inflammation was not reduced in the presence of milk SM (Figure 4D).

Imipramine causes proteolytic degradation of acid sphingomyelinase, which hydrolyzes SM to ceramide and phosphorylcholine. This suggests that a hydrolytic product of SM (e.g., ceramide, sphingosine) is important for milk SM's effect on inflammation.

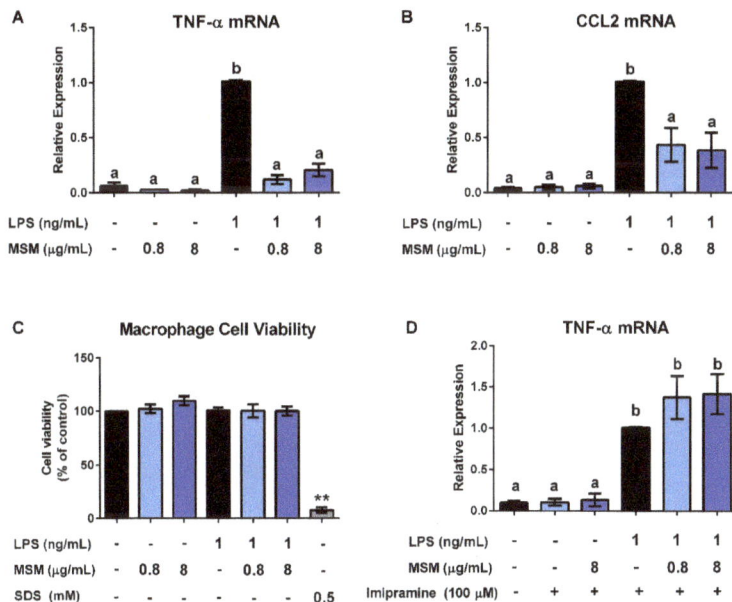

Figure 4. Milk SM inhibits lipopolysaccharide (LPS)-activation of macrophages. RAW264.7 macrophages were pre-incubated with milk SM (MSM) or EtOH vehicle control for 1 h, followed by a 4 h co-incubation ± LPS (1 ng/mL). TNF-α mRNA (**A**) and CCL2 mRNA (**B**) were measured by real-time qRT-PCR. Cell viability determined using Cell Counting Kit-8 (**C**). Sodium dodecyl sulfate (SDS) (0.5 mM) was used on cells as a positive control to induce cell death, ** $p < 0.001$ compared to control. TNF-α mRNA was measured in cells incubated for 2 h with 100 μM imipramine prior to a 1 h incubation with milk SM or control ± a 4 h co-incubation with LPS (**D**). Mean ± SEM, $n = 3$–7 independent experiments. Unlike letters $p < 0.05$ by Holm–Sidak post hoc test.

3.4. Ceramides and Sphingosine, but Not Dihydroceramides, Inhibit LPS Stimulation of Macrophages

Since milk SM contains a mixture of SM species with different fatty acid chain lengths and sphingoid backbones, we tested the effects of C16-ceramide, C24-ceramide, C16-dihydroceramide, and C24-dihydroceramide on macrophage pro-inflammatory gene expression (Figure 5A,B). Interestingly, both C16-ceramide and C24-ceramide at 10 μM concentrations significantly reduced TNF-α mRNA (Figure 5A) and CCL2 mRNA (Figure 5B) in LPS-stimulated macrophages, but their corresponding dihydroceramides did not. None of the ceramides tested altered inflammatory gene expression in the absence of LPS. The anti-inflammatory effects of ceramides (sphingosine base), but not dihydroceramides (sphinganine base), suggest that sphingosine is important for bioactivity. Supporting this notion, sphingosine significantly reduced LPS-stimulation of TNF-α mRNA (Figure 6A) and CCL2 mRNA (Figure 6B) in macrophages. Neither ceramides nor sphingosine significantly affected cell viability (Figure 7A,B) or caspase 3 activity (Figure 7C) at the concentrations tested, suggesting that cell death was not the cause of anti-inflammatory effects.

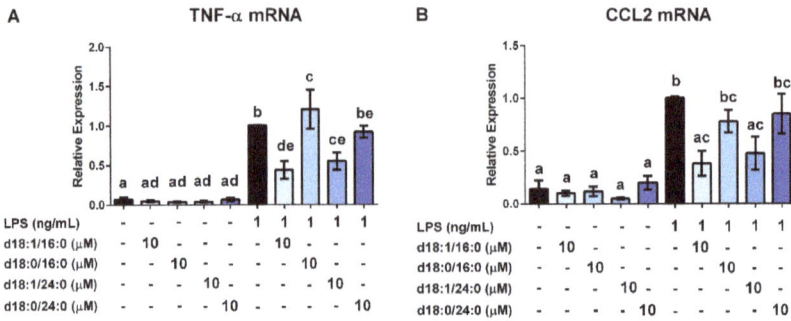

Figure 5. Long-chain ceramides inhibit LPS-activation of macrophages. RAW264.7 macrophages were incubated for 4 h with ceramides (10 μM), dihydroceramides (10 μM), or vehicle control ± LPS (1 ng/mL). C16-ceramide (d18:1/16:0), C24-ceramide (d18:1/24:0), C16-dihydroceramide (d18:0/16:0), and C24-dihydroceramide (d18:0/24:0) were tested. TNF-α mRNA (**A**) and CCL2 mRNA (**B**) measured by real-time qRT-PCR. Mean ± SEM, n = 3–7 independent experiments. Unlike letters $p < 0.05$ by Holm–Sidak post hoc test.

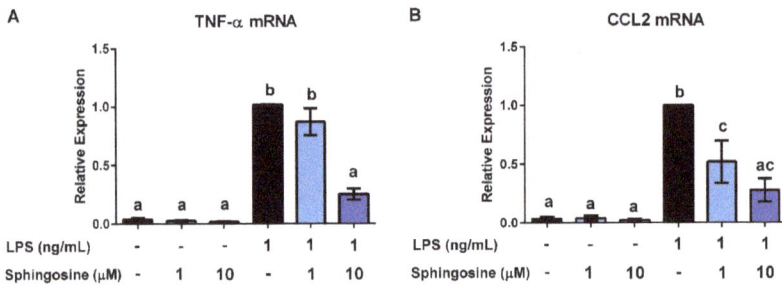

Figure 6. Sphingosine inhibits LPS-activation of macrophages. RAW264.7 macrophages were incubated for 4 h with sphingosine (1–10 μM) or vehicle control ± LPS (1 ng/mL). TNF-α mRNA (**A**) and CCL2 mRNA (**B**) measured by real-time qRT-PCR. Mean ± SEM, n = 3–7 independent experiments. Unlike letters $p < 0.05$ by Holm–Sidak post hoc test.

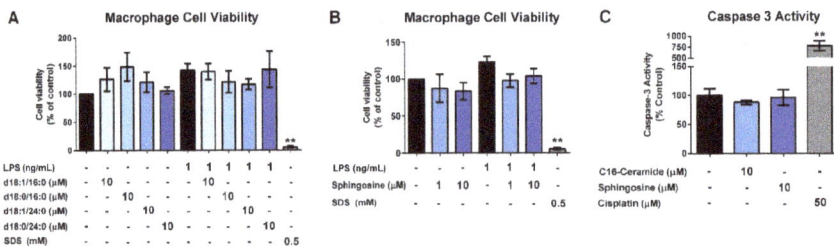

Figure 7. Ceramides and sphingosine do not affect cell viability or apoptosis. Cell viability determined using Cell Counting Kit-8 for ceramides (**A**) and sphingosine (**B**). Sodium dodecyl sulfate (SDS) (0.5 mM) was used on cells as a positive control to induce cell death. Caspase 3 activity measured as an indicator of apoptosis (**C**). Cisplatin (50 μM) was used on cells as a positive control to induce apoptosis. Mean ± SEM, n = 3–5 independent experiments. ** $p < 0.01$ compared to control.

4. Discussion

Chronic low-grade inflammation is a common underlying factor in many diseases afflicting Western societies, including diabetes, non-alcoholic fatty liver disease, and atherosclerosis [39]. Low-grade inflammation in these states may be exacerbated by circulating LPS, also known as endotoxin [2]. Lipopolysaccharide translocation from the gut to circulation is enhanced by lipid absorption [6]. Dietary SM has been studied for its effects on dyslipidemia because it reduces the absorption of other lipids (e.g., fat and cholesterol), and therefore, could potentially influence systemic inflammation. In this study, the addition of 0.1% (w/w) dietary milk SM to a lard-based HFD significantly reduced systemic inflammation markers and tended to lower circulating LPS, but had no effect on gene expression of inflammation or gut barrier function markers in the mesenteric adipose tissue, small intestine, or colon. Furthermore, microbiota composition of cecal feces was mostly unaffected by the addition of milk SM to the HFD, suggesting that gut microbiota compositional differences did not play a major role in altering systemic inflammation markers. Milk SM directly attenuated LPS stimulation of pro-inflammatory gene expression in RAW 264.7 macrophages. Interestingly, hydrolytic products of milk SM (C16-ceramide, C24-ceramide, and sphingosine) showed similar anti-inflammatory effects, whereas the inhibition of SM hydrolysis with imipramine prevented such effects. These results suggest that dietary SM may have additional bioactive effects beyond inhibiting lipid absorption through the lowering of macrophage inflammatory responses, potentially influencing metabolic disease.

In rodent studies, dietary patterns rich in saturated fat and cholesterol, so-called "Western" diets, have been linked to chronic disease through increasing inflammation [40,41]. Detrimental effects of Western diets in mice appear to be influenced by the presence of microbes in the gastrointestinal tract, since depletion of gut microbiota with oral antibiotics reduces the inflammatory response [42]. This is partly related to the effect that Western diets have on increasing LPS in circulation, termed "metabolic endotoxemia", which has been shown to precede increases in systemic inflammation in animal models [2,42,43]. There is also evidence that diets high in fat and cholesterol increase endotoxin concentration in the bloodstream of humans [44,45]. With the current study, we fed milk SM (0.1%) for 10 weeks to HFD-fed mice (60% kcal as fat; 0.15% cholesterol added) to determine if dietary SM could affect chronic low-grade inflammation. Milk SM strongly reduced serum inflammatory cytokines/chemokines and tended to reduce serum LPS (p = 0.08). Lower serum CCL2 and mRNA expression of inflammation markers in epididymal adipose tissue were also previously shown in mice fed milk SM [29]. We have previously reported that the 4-week dietary supplementation of a higher dose of milk SM (0.25% w/w) reduced circulating LPS by 35% in mice fed a milkfat-based high-fat diet [28]. Although we did not observe any differences in mRNA expression of gut barrier markers in the small intestine, reductions in circulating LPS are expected to be partly due to the inhibitory effect of milk SM on lipid absorption. Supporting this notion, we have previously observed reductions in serum and hepatic lipids with the feeding of milk SM [28,29].

Recently, chronic feeding of HFD to mice was shown to increase colonic pro-inflammatory macrophages and gut inflammation, which promoted LPS translocation, insulin resistance, and adipose tissue inflammation [9]. These effects were mediated by early induction of CCL2 in intestinal epithelial cells, which attracted pro-inflammatory macrophages to colon tissue [9]. In this study, although we observed significantly greater colon CCL2 mRNA expression with milk SM intake, we did not observe any significant changes between groups in gene expression of gut barrier markers or macrophage infiltration/inflammation in colon and mesenteric adipose tissues. This suggests that CCL2 mRNA expression was possibly induced with milk SM because of greater dietary fat/cholesterol reaching the colon, but this did not induce macrophage infiltration and TNF-α expression in the colon and surrounding adipose tissue. Although dietary SM is known to inhibit cholesterol absorption, luminal cholesterol can reciprocally inhibit SM digestion [16]. Feeding mice an HFD has also been shown to reduce the expression of the key enzyme in SM digestion, alkaline sphingomyelinase [46,47]. Therefore, it is possible that the use of a high-fat lard diet with added cholesterol in this study influenced ileal and colonic exposure to SM and bioactive metabolic products (ceramide, sphingosine, and sphingosine-1-phosphate), which in turn could affect colon inflammation [48,49].

The intake of HFD has been shown to negatively alter the gut microbiota [50]. Gut dysbiosis has been linked with endotoxin-mediated chronic disease in mouse models [51,52]. Sphingosine, a hydrolytic product of dietary SM, is known to have bactericidal effects [53]. We have previously reported that supplementation of 0.25% (w/w) milk SM to HFD-fed mice altered the fecal microbiota composition by reducing Gram-negative bacterial phyla and increasing *Bifidobacterium* [28]. In the current study, however, we did not observe such large changes in gut microbiota composition, possibly related to the lower dose used or the more severe HFD challenge. We observed a significant difference in the relative abundance of *Acetatifactor* genus, with higher levels in mice fed milk SM. This genus was shown to be isolated from the intestine of obese mice and produces the short-chain fatty acids, acetate and butyrate [54]. The abundance of *Acetatifactor* in the gut has also been shown to be related to lard intake in mice and correlated with the concentrations of secondary bile acids in mouse cecum [55]. We also have observed increases in *Acetatifactor* genus with 0.25% (w/w) milk SM feeding on a low-fat diet background (Norris and Blesso, unpublished observations 2016). At this point, it is unclear if these bacteria contribute to the inflammatory changes observed or in response to the effects of milk SM on lipid absorption. Further research is needed to clarify these associations.

Milk SM significantly reduced TNF-α and CCL2 mRNA expression in LPS-stimulated RAW264.7 macrophages. There are several potential mechanisms that could explain this observed effect. First, there could be a direct interaction between milk SM and LPS within the media, where intact SM itself is neutralizing LPS, possibly due to the formation of a complex. In our current study, this would mean that the LPS would be neutralized by SM before it could interact with the cell, as part of a lipid vesicle that masked the reactive lipid-A moiety of LPS. Incubation with 100 μM bovine brain SM was reported to neutralize LPS stimulation of polymorphonuclear leukocytes (PMN) in the presence of soluble CD14 and LBP [25]. Furthermore, SM hydrolysis can yield bioactive products, including ceramide and sphingosine. Imipramine is a tricyclic anti-depressant (TCA) that causes proteolytic degradation of the enzyme acid-SMase [36]. Imipramine may also have other effects besides acid-SMase degradation, as it was shown that acid ceramidase was down-regulated by another TCA, desipramine [56]. Interestingly, the addition of imipramine completely abolished the anti-inflammatory effects of milk SM. This indicates that one of the hydrolytic products of SM (e.g., ceramide, sphingosine, phosphorylcholine) may be responsible for the reduction in inflammation. If milk SM were hydrolyzed by SMase and ceramidase to ceramide and sphingosine, respectively, each of those products could potentially exert anti-inflammatory effects. Longer-chain ceramides may compete with LPS and block TLR4 signaling, as they have been shown to bind to CD14 in monocytes and form a multi-molecular complex that contains CD36, but lacks TLR4 [57]. To confirm this, ceramides containing different amide-linked fatty acid chain lengths (16:0 and 24:0) and sphingoid bases (dihydroceramides) were tested. We used ceramides with 16-carbon and 24-carbon fatty acid chain lengths because these ceramides would be produced from the hydrolysis of the natural mixture of SM species that comprise milk SM. Additionally, these fatty acid chain lengths within the ceramide have the ability to alter its bioactivity [58]. Specifically, it has been shown that intracellular C16-ceramide is particularly pro-apoptotic, whereas C24-ceramide and dihydroceramides antagonize apoptosis induced by shorter ceramides [59–61]. Interestingly, both C16-ceramide and C24-ceramide treatment of macrophages resulted in a significant reduction in TNF-α and CCL2 mRNA expression, while dihydroceramides had no effect. None of the ceramides or dihydroceramides affected cell viability or the early apoptosis marker, caspase 3 activity, suggesting that the effects were not due to an induction of cell death. Although intracellular ceramides have been shown to be pro-apoptotic [58], exogenously added long-chain ceramides do not affect viability of macrophages [62]. These data indicate that ceramides contribute to the anti-inflammatory effect of milk SM, and strengthen the findings of the imipramine data. However, since only the sphingosine base containing-ceramides showed anti-inflammatory effects and not sphinganine-containing dihydroceramides, we tested the effects of sphingosine on LPS stimulation of macrophages. We observed similar effects of sphingosine (d18:1) on inhibiting TNF-α and CCL2 mRNA without affecting cell viability or apoptosis, as C16- and C24-ceramides. Sphingosine could display anti-inflammatory effects through the activation of the nuclear receptor, peroxisome proliferator-activate receptor-gamma (PPARγ) [22]. In support of our findings, exogenous C8-ceramide and sphingosine have been previously reported to reduce TNF-α secretion from LPS-stimulated macrophages [63]. Furthermore, both C16-ceramide and membrane-permeable C2-ceramide have been shown to reduce TNF-α secretion from LPS-stimulated macrophages by post-translational mechanisms [64].

In summary, dietary milk SM reduced systemic inflammation in diet-induced obese mice, without altering gene expression of macrophage or barrier function markers in the mesenteric adipose tissue or colon. Milk SM directly attenuated LPS stimulation of pro-inflammatory gene expression in RAW 264.7 macrophages. The inhibition of SM hydrolysis prevented milk SM anti-inflammatory effects, while the hydrolytic products C16-ceramide, C24-ceramide, and sphingosine showed anti-inflammatory effects. In addition to effects on lipid absorption, dietary SM may have additional bioactive effects by lowering inflammation and LPS activity, potentially influencing metabolic disease. Further research is warranted to confirm the mechanism and effect of dietary milk SM on inflammation.

Supplementary Materials: The following are available online at http://www.mdpi.com/2306-5710/3/3/37/s1. Table S1: Primer List for qRT-PCR.

Acknowledgments: This work was supported by research funding provided by the University of Connecticut to C.N.B.

Author Contributions: G.H.N. and C.M.P. conducted the research, analyzed data, and wrote the paper; C.J. conducted the research; C.N.B. designed the research, analyzed data, and had primary responsibility for final content. All authors read and approved the final manuscript.

Conflicts of Interest: All authors claim no conflicts of interest.

References

1. Hotamisligil, G.S. Inflammation and metabolic disorders. *Nature* **2006**, *444*, 860–867. [CrossRef] [PubMed]
2. Cani, P.D.; Amar, J.; Iglesias, M.A.; Poggi, M.; Knauf, C.; Bastelica, D.; Neyrinck, A.M.; Fava, F.; Tuohy, K.M.; Chabo, C.; et al. Metabolic endotoxemia initiates obesity and insulin resistance. *Diabetes* **2007**, *56*, 1761–1772. [CrossRef] [PubMed]
3. Lu, Y.C.; Yeh, W.C.; Ohashi, P.S. Lps/tlr4 signal transduction pathway. *Cytokine* **2008**, *42*, 145–151. [CrossRef] [PubMed]
4. Suzuki, T. Regulation of intestinal epithelial permeability by tight junctions. *Cell. Mol. Life Sci.* **2013**, *70*, 631–659. [CrossRef] [PubMed]
5. Neurath, M.F. Cytokines in inflammatory bowel disease. *Nat. Rev. Immunol.* **2014**, *14*, 329–342. [CrossRef] [PubMed]
6. Ghoshal, S.; Witta, J.; Zhong, J.; de Villiers, W.; Eckhardt, E. Chylomicrons promote intestinal absorption of lipopolysaccharides. *J. Lipid Res.* **2009**, *50*, 90–97. [CrossRef] [PubMed]
7. Serino, M.; Luche, E.; Gres, S.; Baylac, A.; Berge, M.; Cenac, C.; Waget, A.; Klopp, P.; Iacovoni, J.; Klopp, C.; et al. Metabolic adaptation to a high-fat diet is associated with a change in the gut microbiota. *Gut* **2012**, *61*, 543–553. [CrossRef] [PubMed]
8. Tilg, H.; Moschen, A.R.; Kaser, A. Obesity and the microbiota. *Gastroenterology* **2009**, *136*, 1476–1483. [CrossRef] [PubMed]
9. Kawano, Y.; Nakae, J.; Watanabe, N.; Kikuchi, T.; Tateya, S.; Tamori, Y.; Kaneko, M.; Abe, T.; Onodera, M.; Itoh, H. Colonic pro-inflammatory macrophages cause insulin resistance in an intestinal ccl2/ccr2-dependent manner. *Cell Metab.* **2016**, *24*, 295–310. [CrossRef] [PubMed]
10. Blesso, C.N. Egg phospholipids and cardiovascular health. *Nutrients* **2015**, *7*, 2731–2747. [CrossRef] [PubMed]
11. Norris, G.H.; Blesso, C.N. Dietary sphingolipids: Potential for management of dyslipidemia and nonalcoholic fatty liver disease. *Nutr. Rev.* **2017**, *75*, 274–285. [CrossRef] [PubMed]

12. Vesper, H.; Schmelz, E.M.; Nikolova-Karakashian, M.N.; Dillehay, D.L.; Lynch, D.V.; Merrill, A.H., Jr. Sphingolipids in food and the emerging importance of sphingolipids to nutrition. *J. Nutr.* **1999**, *129*, 1239–1250. [PubMed]

13. Hannich, J.T.; Umebayashi, K.; Riezman, H. Distribution and functions of sterols and sphingolipids. *Cold Spring Harb. Perspect. Biol.* **2011**, *3*, a004762. [CrossRef] [PubMed]

14. Rombaut, R.; Dewettinck, K. Properties, analysis and purification of milk polar lipids. *Int. Dairy J.* **2006**, *16*, 1362–1373. [CrossRef]

15. Noh, S.K.; Koo, S.I. Egg sphingomyelin lowers the lymphatic absorption of cholesterol and alpha-tocopherol in rats. *J. Nutr.* **2003**, *133*, 3571–3576. [PubMed]

16. Nyberg, L.; Duan, R.D.; Nilsson, A. A mutual inhibitory effect on absorption of sphingomyelin and cholesterol. *J. Nutr. Biochem.* **2000**, *11*, 244–249. [CrossRef]

17. Duivenvoorden, I.; Voshol, P.J.; Rensen, P.C.; van Duyvenvoorde, W.; Romijn, J.A.; Emeis, J.J.; Havekes, L.M.; Nieuwenhuizen, W.F. Dietary sphingolipids lower plasma cholesterol and triacylglycerol and prevent liver steatosis in apoe*3leiden mice. *Am. J. Clin. Nutr.* **2006**, *84*, 312–321. [PubMed]

18. Noh, S.K.; Koo, S.I. Milk sphingomyelin is more effective than egg sphingomyelin in inhibiting intestinal absorption of cholesterol and fat in rats. *J. Nutr.* **2004**, *134*, 2611–2616. [PubMed]

19. Garmy, N.; Taieb, N.; Yahi, N.; Fantini, J. Interaction of cholesterol with sphingosine: Physicochemical characterization and impact on intestinal absorption. *J. Lipid Res.* **2005**, *46*, 36–45. [CrossRef] [PubMed]

20. Eckhardt, E.R.; Wang, D.Q.; Donovan, J.M.; Carey, M.C. Dietary sphingomyelin suppresses intestinal cholesterol absorption by decreasing thermodynamic activity of cholesterol monomers. *Gastroenterology* **2002**, *122*, 948–956. [CrossRef] [PubMed]

21. Chung, R.W.; Kamili, A.; Tandy, S.; Weir, J.M.; Gaire, R.; Wong, G.; Meikle, P.J.; Cohn, J.S.; Rye, K.A. Dietary sphingomyelin lowers hepatic lipid levels and inhibits intestinal cholesterol absorption in high-fat-fed mice. *PLoS ONE* **2013**, *8*, e55949. [CrossRef] [PubMed]

22. Mazzei, J.C.; Zhou, H.; Brayfield, B.P.; Hontecillas, R.; Bassaganya-Riera, J.; Schmelz, E.M. Suppression of intestinal inflammation and inflammation-driven colon cancer in mice by dietary sphingomyelin: Importance of peroxisome proliferator-activated receptor gamma expression. *J. Nutr. Biochem.* **2011**, *22*, 1160–1171. [CrossRef] [PubMed]

23. Fischbeck, A.; Leucht, K.; Frey-Wagner, I.; Bentz, S.; Pesch, T.; Kellermeier, S.; Krebs, M.; Fried, M.; Rogler, G.; Hausmann, M.; Humpf, H.U. Sphingomyelin induces cathepsin D-mediated apoptosis in intestinal epithelial cells and increases inflammation in DSS colitis. *Gut* **2011**, *60*, 55–65. [CrossRef] [PubMed]

24. Leucht, K.; Fischbeck, A.; Caj, M.; Liebisch, G.; Hartlieb, E.; Benes, P.; Fried, M.; Humpf, H.U.; Rogler, G.; Hausmann, M. Sphingomyelin and phosphatidylcholine contrarily affect the induction of apoptosis in intestinal epithelial cells. *Mol. Nutr. Food Res.* **2014**, *58*, 782–798. [CrossRef] [PubMed]

25. Wurfel, M.M.; Wright, S.D. Lipopolysaccharide-binding protein and soluble cd14 transfer lipopolysaccharide to phospholipid bilayers: Preferential interaction with particular classes of lipid. *J. Immunol.* **1997**, *158*, 3925–3934. [PubMed]

26. Parker, T.S.; Levine, D.M.; Chang, J.C.; Laxer, J.; Coffin, C.C.; Rubin, A.L. Reconstituted high-density lipoprotein neutralizes gram-negative bacterial lipopolysaccharides in human whole blood. *Infect. Immun.* **1995**, *63*, 253–258. [PubMed]

27. Memon, R.A.; Holleran, W.M.; Moser, A.H.; Seki, T.; Uchida, Y.; Fuller, J.; Shigenaga, J.K.; Grunfeld, C.; Feingold, K.R. Endotoxin and cytokines increase hepatic sphingolipid biosynthesis and produce lipoproteins enriched in ceramides and sphingomyelin. *Arterioscler. Thromb. Vasc. Biol.* **1998**, *18*, 1257–1265. [CrossRef] [PubMed]

28. Norris, G.H.; Jiang, C.; Ryan, J.; Porter, C.M.; Blesso, C.N. Milk sphingomyelin improves lipid metabolism and alters gut microbiota in high fat diet-fed mice. *J. Nutr. Biochem.* **2016**, *30*, 93–101. [CrossRef] [PubMed]

29. Norris, G.H.; Porter, C.M.; Jiang, C.; Millar, C.L.; Blesso, C.N. Dietary sphingomyelin attenuates hepatic steatosis and adipose tissue inflammation in high-fat-diet-induced obese mice. *J. Nutr. Biochem.* **2017**, *40*, 36–43. [CrossRef] [PubMed]

30. Nair, A.B.; Jacob, S. A simple practice guide for dose conversion between animals and human. *J. Basic Clin. Pharm.* **2016**, *7*, 27–31. [CrossRef] [PubMed]

31. Caporaso, J.G.; Lauber, C.L.; Walters, W.A.; Berg-Lyons, D.; Huntley, J.; Fierer, N.; Owens, S.M.; Betley, J.; Fraser, L.; Bauer, M.; et al. Ultra-high-throughput microbial community analysis on the illumina hiseq and miseq platforms. *ISME J.* **2012**, *6*, 1621–1624. [CrossRef] [PubMed]

32. Kozich, J.J.; Westcott, S.L.; Baxter, N.T.; Highlander, S.K.; Schloss, P.D. Development of a dual-index sequencing strategy and curation pipeline for analyzing amplicon sequence data on the miseq illumina sequencing platform. *Appl. Environ. Microbiol.* **2013**, *79*, 5112–5120. [CrossRef] [PubMed]

33. Bioinformatics/mothur.batch. Available online: https://github.com/krmaas/bioinformatics/blob/master/mothur.batch (accessed on 18 July 2017).

34. Quast, C.; Pruesse, E.; Yilmaz, P.; Gerken, J.; Schweer, T.; Yarza, P.; Peplies, J.; Glockner, F.O. The silva ribosomal rna gene database project: Improved data processing and web-based tools. *Nucleic Acids Res.* **2013**, *41*, D590–596. [CrossRef] [PubMed]

35. Wang, Q.; Garrity, G.M.; Tiedje, J.M.; Cole, J.R. Naive bayesian classifier for rapid assignment of rrna sequences into the new bacterial taxonomy. *Appl. Environ. Microbiol.* **2007**, *73*, 5261–5267. [CrossRef] [PubMed]

36. Hurwitz, R.; Ferlinz, K.; Sandhoff, K. The tricyclic antidepressant desipramine causes proteolytic degradation of lysosomal sphingomyelinase in human fibroblasts. *Biol. Chem. Hoppe Seyler* **1994**, *375*, 447–450. [CrossRef] [PubMed]

37. De Caceres, M.; Legendre, P. Associations between species and groups of sites: Indices and statistical inference. *Ecology* **2009**, *90*, 3566–3574. [CrossRef] [PubMed]

38. Alrefai, W.A.; Annaba, F.; Sarwar, Z.; Dwivedi, A.; Saksena, S.; Singla, A.; Dudeja, P.K.; Gill, R.K. Modulation of human niemann-pick c1-like 1 gene expression by sterol: Role of sterol regulatory element binding protein 2. *Am. J. Physiol. Gastrointest. Liver Physiol.* **2007**, *292*, G369–G376. [CrossRef] [PubMed]

39. Baker, R.G.; Hayden, M.S.; Ghosh, S. Nf-kappab, inflammation, and metabolic disease. *Cell Metab.* **2011**, *13*, 11–22. [CrossRef] [PubMed]

40. Wouters, K.; van Gorp, P.J.; Bieghs, V.; Gijbels, M.J.; Duimel, H.; Lutjohann, D.; Kerksiek, A.; van Kruchten, R.; Maeda, N.; Staels, B.; et al. Dietary cholesterol, rather than liver steatosis, leads to hepatic inflammation in hyperlipidemic mouse models of nonalcoholic steatohepatitis. *Hepatology* **2008**, *48*, 474–486. [CrossRef] [PubMed]

41. Wang, Y.; Qian, Y.; Fang, Q.; Zhong, P.; Li, W.; Wang, L.; Fu, W.; Zhang, Y.; Xu, Z.; Li, X.; et al. Saturated palmitic acid induces myocardial inflammatory injuries through direct binding to tlr4 accessory protein md2. *Nat. Commun.* **2017**, *8*, 13997. [CrossRef] [PubMed]

42. Cani, P.D.; Bibiloni, R.; Knauf, C.; Waget, A.; Neyrinck, A.M.; Delzenne, N.M.; Burcelin, R. Changes in gut microbiota control metabolic endotoxemia-induced inflammation in high-fat diet-induced obesity and diabetes in mice. *Diabetes* **2008**, *57*, 1470–1481. [CrossRef] [PubMed]

43. Ding, S.; Chi, M.M.; Scull, B.P.; Rigby, R.; Schwerbrock, N.M.; Magness, S.; Jobin, C.; Lund, P.K. High-fat diet: Bacteria interactions promote intestinal inflammation which precedes and correlates with obesity and insulin resistance in mouse. *PLoS ONE* **2010**, *5*, e12191. [CrossRef] [PubMed]

44. Pendyala, S.; Walker, J.M.; Holt, P.R. A high-fat diet is associated with endotoxemia that originates from the gut. *Gastroenterology* **2012**, *142*, 1100–1101. [CrossRef] [PubMed]

45. Erridge, C.; Attina, T.; Spickett, C.M.; Webb, D.J. A high-fat meal induces low-grade endotoxemia: Evidence of a novel mechanism of postprandial inflammation. *Am. J. Clin. Nutr.* **2007**, *86*, 1286–1292. [PubMed]

46. Zhang, Y.; Cheng, Y.; Hansen, G.H.; Niels-Christiansen, L.L.; Koentgen, F.; Ohlsson, L.; Nilsson, A.; Duan, R.D. Crucial role of alkaline sphingomyelinase in sphingomyelin digestion: A study on enzyme knockout mice. *J. Lipid Res.* **2011**, *52*, 771–781. [CrossRef] [PubMed]

47. Cheng, Y.; Ohlsson, L.; Duan, R.D. Psyllium and fat in diets differentially affect the activities and expressions of colonic sphingomyelinases and caspase in mice. *Br. J. Nutr.* **2004**, *91*, 715–723. [CrossRef] [PubMed]

48. Nagahashi, M.; Hait, N.C.; Maceyka, M.; Avni, D.; Takabe, K.; Milstien, S.; Spiegel, S. Sphingosine-1-phosphate in chronic intestinal inflammation and cancer. *Adv. Biol. Regul.* **2014**, *54*, 112–120. [CrossRef] [PubMed]

49. Nilsson, Å. Role of sphingolipids in infant gut health and immunity. *J. Pediatr.* **2016**, *173*, S53–S59. [CrossRef] [PubMed]

50. Murphy, E.F.; Cotter, P.D.; Healy, S.; Marques, T.M.; O'Sullivan, O.; Fouhy, F.; Clarke, S.F.; O'Toole, P.W.; Quigley, E.M.; Stanton, C.; et al. Composition and energy harvesting capacity of the gut microbiota: Relationship to diet, obesity and time in mouse models. *Gut* **2010**, *59*, 1635–1642. [CrossRef] [PubMed]

51. Henao-Mejia, J.; Elinav, E.; Jin, C.; Hao, L.; Mehal, W.Z.; Strowig, T.; Thaiss, C.A.; Kau, A.L.; Eisenbarth, S.C.; Jurczak, M.J.; et al. Inflammasome-mediated dysbiosis regulates progression of nafld and obesity. *Nature* **2012**, *482*, 179–185. [CrossRef] [PubMed]

52. De Minicis, S.; Rychlicki, C.; Agostinelli, L.; Saccomanno, S.; Candelaresi, C.; Trozzi, L.; Mingarelli, E.; Facinelli, B.; Magi, G.; Palmieri, C.; et al. Dysbiosis contributes to fibrogenesis in the course of chronic liver injury in mice. *Hepatology* **2014**, *59*, 1738–1749. [CrossRef] [PubMed]

53. Sprong, R.C.; Hulstein, M.F.; Van der Meer, R. Bactericidal activities of milk lipids. *Antimicrob. Agents Chemother.* **2001**, *45*, 1298–1301. [CrossRef] [PubMed]

54. Pfeiffer, N.; Desmarchelier, C.; Blaut, M.; Daniel, H.; Haller, D.; Clavel, T. Acetatifactor muris gen. Nov., sp. Nov., a novel bacterium isolated from the intestine of an obese mouse. *Arch. Microbiol.* **2012**, *194*, 901–907. [CrossRef] [PubMed]

55. Kubeck, R.; Bonet-Ripoll, C.; Hoffmann, C.; Walker, A.; Muller, V.M.; Schuppel, V.L.; Lagkouvardos, I.; Scholz, B.; Engel, K.H.; Daniel, H.; et al. Dietary fat and gut microbiota interactions determine diet-induced obesity in mice. *Mol. Metab.* **2016**, *5*, 1162–1174. [CrossRef] [PubMed]

56. Zeidan, Y.H.; Pettus, B.J.; Elojeimy, S.; Taha, T.; Obeid, L.M.; Kawamori, T.; Norris, J.S.; Hannun, Y.A. Acid ceramidase but not acid sphingomyelinase is required for tumor necrosis factor-{alpha}-induced pge2 production. *J. Biol. Chem.* **2006**, *281*, 24695–24703. [CrossRef] [PubMed]

57. Pfeiffer, A.; Bottcher, A.; Orso, E.; Kapinsky, M.; Nagy, P.; Bodnar, A.; Spreitzer, I.; Liebisch, G.; Drobnik, W.; Gempel, K.; et al. Lipopolysaccharide and ceramide docking to cd14 provokes ligand-specific receptor clustering in rafts. *Eur. J. Immunol.* **2001**, *31*, 3153–3164. [CrossRef]

58. Grosch, S.; Schiffmann, S.; Geissslinger, G. Chain length-specific properties of ceramides. *Prog. Lipid Res.* **2012**, *51*, 50–62. [CrossRef] [PubMed]

59. Bielawska, A.; Crane, H.M.; Liotta, D.; Obeid, L.M.; Hannun, Y.A. Selectivity of ceramide-mediated biology. Lack of activity of erythro-dihydroceramide. *J. Biol. Chem.* **1993**, *268*, 26226–26232. [PubMed]

60. Stiban, J.; Fistere, D.; Colombini, M. Dihydroceramide hinders ceramide channel formation: Implications on apoptosis. *Apoptosis* **2006**, *11*, 773–780. [CrossRef] [PubMed]

61. Stiban, J.; Perera, M. Very long chain ceramides interfere with c16-ceramide-induced channel formation: A plausible mechanism for regulating the initiation of intrinsic apoptosis. *Biochim. Biophys. Acta* **2015**, *1848*, 561–567. [CrossRef] [PubMed]

62. Shabbits, J.A.; Mayer, L.D. Intracellular delivery of ceramide lipids via liposomes enhances apoptosis in vitro. *Biochim. Biophys. Acta* **2003**, *1612*, 98–106. [CrossRef]

63. Jozefowski, S.; Czerkies, M.; Lukasik, A.; Bielawska, A.; Bielawski, J.; Kwiatkowska, K.; Sobota, A. Ceramide and ceramide 1-phosphate are negative regulators of tnf-alpha production induced by lipopolysaccharide. *J. Immunol.* **2010**, *185*, 6960–6973. [CrossRef] [PubMed]

64. Rozenova, K.A.; Deevska, G.M.; Karakashian, A.A.; Nikolova-Karakashian, M.N. Studies on the role of acid sphingomyelinase and ceramide in the regulation of tumor necrosis factor alpha (tnfalpha)-converting enzyme activity and tnfalpha secretion in macrophages. *J. Biol. Chem.* **2010**, *285*, 21103–21113. [CrossRef] [PubMed]

beverages

MDPI

Short Note

Organic vs. Conventional Milk: Some Considerations on Fat-Soluble Vitamins and Iodine Content

Pamela Manzi * and Alessandra Durazzo

Consiglio per la ricerca in agricoltura e l'analisi dell'economia agraria—Centro di ricerca Alimenti e Nutrizione (CREA-AN), Via Ardeatina 546, 00178 Roma, Italy; alessandra.durazzo@crea.gov.it
* Correspondence: pamela.manzi@crea.gov.it; Tel.: +39-065-149-4499; Fax: +39-065-149-550

Academic Editor: Edgar Chambers IV
Received: 30 May 2017; Accepted: 27 July 2017; Published: 1 August 2017

Abstract: The organic food market is considerably expanding all over the world, and the related dairy market represents its third most important sector. The reason lies in the fact that consumers tend to associate organic dairy products with positive perceptions: organic milk is eco- and animal-friendly, is not produced with antibiotics or hormones, and according to general opinion, provides additional nutrients and beneficial properties. These factors justify its higher cost. These are the reasons that explain extensive research into the comparison of the differences in the amount of chemical compounds between organic and conventional milk. However, it is not simple to ascertain the potential advantage of organic food from the nutritional point of view, because this aspect should be determined within the context of the total diet. Thus, considering all the factors described above, the purpose of this work is to compare the amount of selected nutrients (i.e., iodine and the fat-soluble vitamins such as alfa-tocopherol and beta-carotene) in organic and conventional milk, expressed as the percentage of recommended daily intakes in one serving. In detail, in order to establish the real share of these biologically active compounds to the total diet, their percent contribution was calculated using the Dietary Reference Values for adults (both men and women) adopted by the European Food Safety Authority. According to these preliminary considerations, the higher cost of organic milk can mainly be explained by the high costs of the management of specific farms and no remarkable or substantial benefits in human health can be ascribed to the consumption of organic milk. In this respect, this paper wants to make a small contribution to the estimation of the potential value and nutritional health benefits of organic food, even though further studies are needed.

Keywords: organic milk; conventional milk; dietary assessment; chemical components

1. Introduction

Over the last decade, the organic market has broadly expanded [1]. The main world leaders in organic food are North America and Europe, and according to the projections, these food items will grow by 40% in the next decade [2]. In addition to the market of organic fruit and vegetables, the sectors of organic milk and meat have also developed extensively over the last decade. Sales of organic milk increased between 2006 and 2013 (with Germany and France leading the growth), and in the UK, organic milk consumption increased (by about 3%) in 2016, while the intake of conventional milk decreased (by about −1.9%) during the same period [3].

Consumers usually associate organic farming methods with ecological practices; i.e., attention to biodiversity, soil quality, and animal welfare, as well as low levels of pesticide residues. However, such techniques are expensive. Consequently, the final products have higher costs, which remain as the main barrier to purchase [4]. Generally, consumers have a positive perception of organic milk because—according to them—it is environmentally- and animal-friendly, it is not produced using antibiotics or hormones, and it is conceived as a source of additional nutrients or beneficial properties.

All these factors justify its higher cost. Nevertheless, the role of the cost of these products in the purchase decision is still a matter of debate.

As mentioned above, one of the reasons why the demand for organic food is now prevalent is the attention to environmental and animal welfare. In particular, it has been observed that local breeds of animals can adapt to the local environment and are more resistant to many diseases, even if organic farming does not protect dairy cows from mastitis [5,6]. In the study of Mueller et al. [7], the potential beneficial impacts of some organic dairy farms on the environment are analysed and compared, considering several environmental impact categories (such as energy consumption, climate impact, land demand, conservation of soil fertility, biodiversity, animal welfare, and milk quality). According to the results of this study, low-input farming had positive effects on animal welfare and milk quality. However, one of the main complications affecting organic farming concerns milk yield. Most studies available today [8–10] highlight that organic milk yield per animal is, on average, lower (by about 20%) than the yield derived from conventional milk.

In this context, this work has been developed to compare the amount of selected nutrients—i.e., fat-soluble vitamins (alfa-tocopherol and beta-carotene) and iodine—in organic and conventional milk, expressed as the percentage of recommended daily intakes in one serving. In detail, the percent contribution of these biologically active compounds both in organic and in conventional milk was calculated using the Dietary Reference Values for adults (men and women) adopted by the European Food Safety Authority, in order to establish the real contribution of the different compounds to the total diet.

2. Organic vs. Conventional Milk: Focus on Fat Fraction and Mineral Content

Many studies about the different concentrations of chemical compounds in organic and conventional milk have been conducted over the last years. Some authors [11] have highlighted how important it is that all factors influencing milk composition (with the exception of the farming system—organic vs. conventional) should be identical in research that evaluates the differences between organic and conventional milk.

The papers on milk proteins and carbohydrates are often contradictory [11], several studies agree on the fatty acid composition, because its proportion easily varies in milk along with changes in diet. Capuano et al. [12] showed some differences in the profiles of fatty acid and triglycerides s in three different kinds of milk: organic, conventional, and milk coming from grazing cows (at least 120 days per year and at least 6 h per day, labelled as "weidemelk"). The results of this study confirmed the presence of significant differences in the profile of fat components between conventional and organic milk. Moreover, according to the authors [12], the fatty acid profile can be used—to a certain extent—to certify organic milk to be sold.

From a nutritional point of view, according to some studies, organic milk contains less omega-6 fatty acids and more omega-3 fatty acids than conventional milk [13,14]. In detail, organic milk shows higher amounts of alpha-linolenic acid, eicosapentaenoic acid, and docosapentaenoic acid than conventional milk, even if some seasonal variability can occur. Likewise, some authors [15] observed similar results in conventional and organic milk produced in Poland. The authors [15] found higher contents of polyunsaturated and omega-3 fatty acids in organic milk than in conventional samples.

Among the fat components of milk, conjugated linoleic acid (CLA)—positional and geometric isomers of linoleic acid—have been the most studied compounds over the last years. CLAs are characterized by conjugated double bonds located at positions 8 and 10, 9 and 11, 10 and 12, or 11 and 13. In milk, the most abundant isomer is *cis*-9,*trans*-11 (rumenic acid, 75–90% of total CLAs). Numerous biological effects have been reported for CLAs: rumenic acid mainly has an anticarcinogenic property, while *trans*-10,*cis*-12 isomer is able to reduce body fat and influence blood lipids [16,17].

As reported by several papers, organic milk has a fatty acid profile richer in alpha-linolenic acid and isomers of conjugated linoleic acid [13,18] than milk from conventional farms. This difference can probably be attributed to the composition of dairy diets (grazing-based diets), with special reference to the different amounts of fresh forage and concentrates that are consumed. Milk from ruminants that are pasture-fed contains more *cis-9,trans-11*-CLA than milk from ruminants fed indoors, although some variations may occur depending on several factors: the diet, the individual parameters of cows, the stage of lactation, the feeding system, or the seasonal effects [19–22]. According to Butler et al. [13], CLA (as C18:2 *cis-9,trans-11*) present in milk from conventional (high-input) and organic (low-input) dairy production systems varies from 8.8 to 14.1 $g \cdot kg^{-1}$ milk fat, respectively, during the outdoor fresh forage-based feeding period. However, this difference is reduced (from 6.2 to 7.8 $g \cdot kg^{-1}$ milk fat) during the indoor (conserved forage-based) feeding period.

After investigating the fatty acid composition and the fat-soluble vitamin contents in organic and conventional Italian dairy products, some authors [23] concluded that among the several parameters, the *cis-9,trans-11* C18:2 (CLA conjugated linoleic acids)/linolenic acid (LA) ratio value characterized better fat in organic than in conventional milk. In addition, this ratio can be used as a marker for the identification of organic dairy products.

It is interesting to mention the meta-analysis of Palupi et al. [24], which summarizes the results of 29 studies on milk fat. According to this study [24], organic dairy products contain greater amounts of ALA (alpha-linolenic acid), omega-3 fatty acid, *cis-9*, *trans-11* conjugated linoleic acid, *trans-11* vaccenic acid, eicosapentaenoic acid, and docosapentaenoic acid than conventional types. The same authors also observed that the former have significantly higher quantities of omega-3 to omega-6 ratio and Δ9-desaturase index than the latter. Likewise, the recent meta-analyses of [25], based on 170 published studies, confirmed that organic milk contains higher n-3 polyunsaturated fatty acids (PUFAs) and conjugated linoleic acid.

Some authors [14] studied omega-6/omega-3 ratios in the U.S. National Organic Program on fatty acids in order to verify the differences in the fatty acid content between organic and conventional milk. In this study [14], this ratio was higher in conventional (5.8) than in organic milk (2.3). Today, Western dietary patterns are characterized by a large quantity of omega-6 with a very high ratio of omega-6/omega-3 (15:1 or 20:1) instead of 1:1 [26], and the reduction of this ratio provides benefits to human health.

Concerning fat-soluble vitamin contents, a previous work by Manzi et al. [27] monitored the nutritional composition of milk obtained from four types of farming: (1) intensive farming with silage; (2) intensive farming with hay concentrate; (3) intensive farming with limited integration of hay concentrate; (4) only pasture. The amount of alpha-tocopherol and beta-carotene reached the highest value in Type 4 farming; a similar performance was detected in the Degree of Antioxidant Protection (DAP) index, calculated as a molar ratio between antioxidant compounds and an oxidation target [28].

A reworking of data by Manzi et al. [27] is reported in Figure 1. The principal component analysis (PCA) performed on analytical data concerning fat-soluble compounds of milk samples derived from four types of farming allowed to identify a partial separation between types 1-2-3 and type 4 (Principal Component 1: 47.6%; Principal Component 2: 27.6%). The loading analysis on the main components made it possible to identify the contribution of the original variables (that is, the analytical determinations that have been carried out) to the model of the principal components with special reference to alpha-tocopherol, beta-carotene, and DAP.

On this subject, it is worth mentioning a similar result described in a recent work of Puppel et al. [29], who applied the same index (DAP) to determine the nutritional value of milk produced in Poland during the indoor feeding season, the pasture feeding season, and the pasture feeding season with added corn grain. The results of Puppel et al. [29] proved that the content of fat-soluble vitamins (vitamin E, A, and beta-carotene) showed the highest values in milk from cows fed with pasture and pasture with added corn grain: as a result, the DAP index and the total antioxidant status (TAS) values were highest in these milks. Likewise, some authors [13,30] verified that concentrations of fat-soluble vitamins (alpha-tocopherol and carotenoids) were considerably higher in milk from low-input farming systems than in milk from high-input systems.

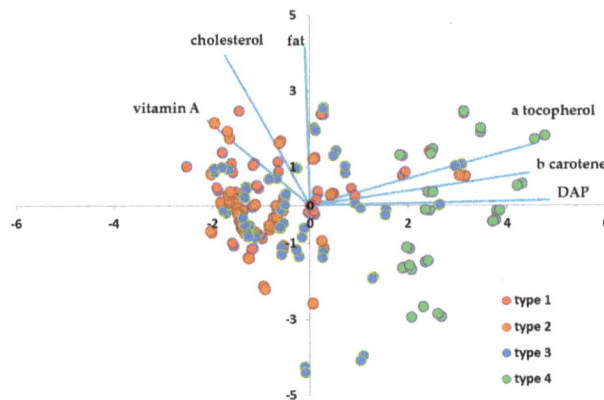

Figure 1. Principal component analysis (PCA) of milk samples from four types of farming: (1) intensive farming with silage; (2) intensive farming with hay concentrate; (3) intensive with limited integration of hay concentrate; (4) only pasture (modified from [27]). DAP: Degree of Antioxidant Protection index.

Generally, milk is considered to be the main source of minerals [31] in human nourishment. Therefore, a focus on the mineral content in organic milk is necessary. Whereas in conventional milk minerals come from concentrate feeds, in organic milk they derive mainly from soil—in particular, from the different pastures. This is a difference that explains why organic milk can display trace elements deficiencies. However, the study of Średnicka-Tober et al. [25] showed that the concentration of the main minerals (such as Ca, Mg, P, and K) does not vary substantially between organic and conventional milk, because the difference due to the farm management is relatively low [32]. Nevertheless, conventional milk showed higher concentrations of iodine and selenium than organic milk.

Among all minerals, iodine concentrations should particularly be monitored: human beings need to absorb trace elements from food, but their diets very often lack these substances. This is also the case for iodine. Iodine is essential for the synthesis of thyroid hormones, and its deficiency leads to hypothyroidism (goitre), irreversible brain damage, or mental retardation. Milk and dairy foods are among the main dietary sources [33] of iodine. The iodine content of milk is influenced by many factors: the contents of iodine in forages, the goitrogens in animal feeds (plants belonging to the cruciferous family or soybean, beet pulp, millet, linseed, etc.), the farm management (organic or conventional).

Concerning iodine contents, it has been observed that in organic milk they are present in concentrations of about 30–40% lower than in conventional milk [34,35]. In this regard, it is worth mentioning the study of Rey-Crespo et al. [36] on essential trace elements and some heavy metal residues in organic and conventional milk carried out in Spain, 2013. This study showed how some minerals (e.g., copper, zinc, iodine, and selenium) are higher in conventional milk than in organic milk because of their presence in the concentrate feeds [36].

3. Nutritional Evaluation of Organic Milk in the Diet: Some Considerations on Fat-Soluble Vitamins and Iodine

Generally, as highlighted by Givens and Lovegrove [37], it is important to evaluate the different nutritional values of organic and conventional milk in the context of the total diet; the same authors have reported that the supply of milk fatty acids in organic systems compared to conventional ones is extremely low when examined in the context of total diets. On the basis of data obtained from the meta-analysis of Średnicka-Tober et al. [25], they [37] calculated the real amount of PUFA in the context of the whole diet, and estimated that the percent variation between conventional and organic milk was 0.44 for polyunsaturated fatty acids, −0.04 for omega-6 fatty acids, and 1.86 for omega-3 fatty acids.

Taking into account the considerations above, in this work, the percent contribution of some active compounds (alpha-tocopherol and beta-carotene) and iodine content from one serving (125 g, according to the Italian dietary guidelines [38]) in both organic milk and conventional milk was calculated using the Dietary Reference Values adopted by the European Food Safety Authority (EFSA).

In order to perform the dietary assessment, the fat-soluble vitamin contents of organic and conventional milk obtained from two different studies [27,39] were considered. For the purpose of these works, organic milk was collected during spring and summer seasons from Apulian dairy farms (a region of southern Italy) [27], while different brands of conventional milk were collected from Italian grocery stores and supermarkets [39].

In Figure 2, the results of the real percent contribution of one serving to the requirement of vitamin E and beta-carotene (data from [27,39] for organic and conventional milk, respectively) are shown.

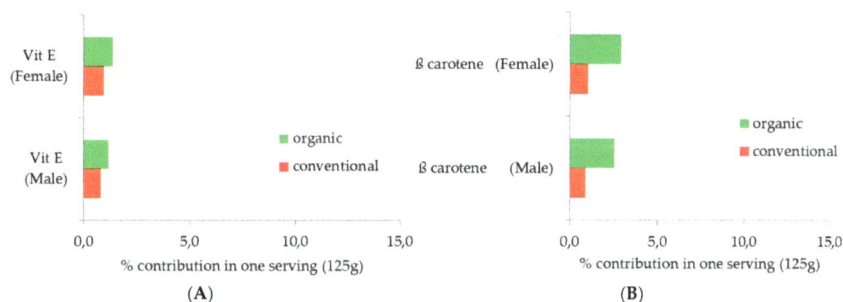

Figure 2. (**A**) Percent contribution in one serving (125 g) of milk to Adequate Intake for vitamin E; (**B**) Percent contribution of beta-carotene to the Population Reference Intake for vitamin A (data from [27,39] were used to achieve the estimate).

In particular, according to EFSA, the Population Reference Intakes (PRIs) for vitamin A are 750 μg RE/day for men and 650 μg RE/day for women [40]. Hence, the difference between organic and conventional milk concerning the contribution of beta-carotene to vitamin A for the recommended daily requirement was not consistent, even though there is a significant dissimilarity ($p < 0.05$) between organic and conventional milk (both in males and females).

According to the EFSA [41], the panel of experts deemed an Adequate Intake of vitamin E (as alpha-tocopherol) in healthy populations to be 13 mg/day for men and 11 mg/day for women. As a result, the differences between organic and conventional milk ($p < 0.05$) concerning the percent contribution of vitamin E for one serving were negligible but significant both in males and females, as reported in Figure 2.

Finally, the evaluation of the percent contribution of iodine in one serving (125 g) was considered. To obtain this information, the Population Reference Intake (150 μg/day for adults from 18 to >75 aged) of EFSA Panel was adopted [42].

The iodine content was obtained from two different studies [34,36]. For the study of Bath et al. [34], both conventional and organic milk was collected from supermarkets in the UK; while, for the study of Rey-Crespo et al. [36], organic milk was collected from dairy farms and conventional milk was collected both from dairy farms and supermarkets in Northern Spain. The percent coverage of iodine for one serving obtained from these researches [34,36] is reported in Figure 3. The results showed that the percent coverage of iodine was lower in organic (6.3% [36] and 12.7% [34]) than in conventional milk (17.1% [36] and 21.4% [34]).

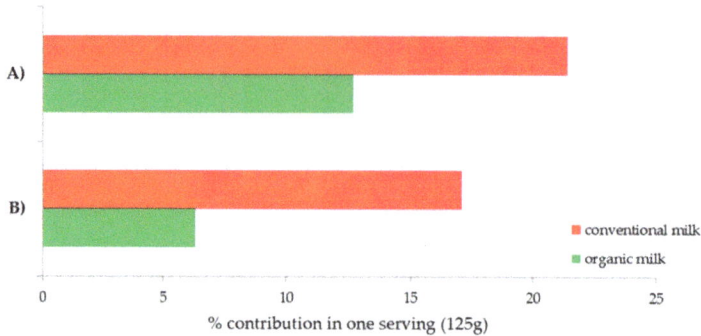

Figure 3. Percent contribution in one serving (125 g) to Population Reference Intake for iodine. (**A**) Data from [34] and (**B**) data from [36] were respectively used to achieve the estimate.

This work has concentrated on fat-soluble vitamins and iodine for two main reasons: 1. there are many studies in the literature concerning these molecules present in conventional and organic milk; 2. fat-soluble vitamins and iodine represent two examples of how organic and conventional milk provide different amounts of certain nutritious substances.

Concerning fat-soluble vitamins, it is important to note that dairy products are not the main source of vitamin E (present in large amounts in vegetable oils such as olive oil and oil seeds), while they are considered a good source of vitamin A. In particular, these fat-soluble vitamins present in organic milk are positively correlated with the animal feed, and the highest concentrations are found in grass, legumes, and other green plants [30,43]. Hence, as highlighted by many studies, organic milk is richer in alpha tocopherol and beta carotene (only present in cow's milk) than conventional milk. It seems necessary to assess whether a higher content of these molecules could somehow bring a considerable benefit to human health [44]. From these preliminary data for fat-soluble vitamins reported in this study and obtained from this nutritional evaluation, no clear indication emerges for recommending organic milk.

At the same time, regarding iodine, the evaluation of its nutritional contribution shows that organic milk cannot provide a satisfactory daily intake. Several factors influence the iodine content of cow's milk, such as the level of iodine supplements in feed, some iodine antagonists (such as glucosinolates) in the feed, or the farm management. The frequent practice of using mineral mixtures containing iodine in conventional farming is one of the reasons for the higher iodine content in conventional milk [45].

Iodine is a central element in human nutrition. It is essential in the regulation of thyroid hormones for cellular metabolism, as well as for the normal growth and development of the body. In this perspective, milk provides significant supplies of this chemical element. These considerations are not intended to convey the idea that the consumption of organic milk is ineffectual—they may motivate farmers to improve their products and the composition of animal feeds. They represent a suggestion to help choose a correct farming system design and maintain a high standard in animal welfare, in order to increase beneficial effects for human beings.

4. Conclusions

Nowadays, consumers are much more attentive to food quality and safety, and they are aware that organic food could meet these needs while restricting the usage of antibiotics and hormones. Hence, researchers' attention should focus on how the consumption of organic food brings benefits to human health.

Establishing whether there is a potential nutritional superiority of organic food is not simple, and there are increasing numbers of reviews available on this subject [46,47]. However, it is not simple to

Beverages **2017**, *3*, 39

conduct epidemiological or intervention studies, because individuals that consume organic foods and dairy products usually have different lifestyles that should be taken into consideration in the definition of these kinds of researches. As recently reported by some authors [48], there are no long-term cohort studies or controlled dietary intervention studies comparing the effects of diets based on organic or conventional food. In this context, the higher cost of organic milk can be justified only by the high cost of the management of these farms.

Moreover, it is worthwhile mentioning that, according to some authors [49], organic farming practices generally have positive impacts on the environment. However, this consideration should not be generalized, because there is a wide and varied range of organic and conventional farms [49].

From a nutritional point of view, according to these preliminary results, no remarkable or substantial benefits to human health can be ascribed to the consumption of organic milk. In this respect, this paper aims to make a small contribution to the estimation of the potential value and nutritional health benefits of organic food, even though further studies are needed. Within this context, this study wants to point out that for an appropriate approach to the nutritional exploitation of organic milk, the evaluation of the real contribution of the different compounds of organic food to the diet should represent the first step.

Acknowledgments: This work was undertaken within the project QUALIFU, financed by the Italian Ministry of Agriculture, Food and Forestry. The authors thank Annalisa Lista for linguistic revision and editing of the manuscript.

Author Contributions: All authors contributed equally to this work: Pamela Manzi has contributed towards data collection, drafting and compiling of the manuscript; Alessandra Durazzo has critically reviewed the manuscript for its accuracy and completeness. All authors read and approved the final manuscript.

Conflicts of Interest: The authors declare no conflict of interest.

References

1. Willer, H.; Yussefi, M.; Sorensen, N. *The World of Organic Agriculture: Statistics and Emerging Trends*; IFOAM: Bonn, Germany; FiBL: Frick, Switzerland, 2009.
2. Costa, S.; Zepeda, L.; Sirieix, L. Exploring the social value of organic food: A qualitative study in france. *Int. J. Consum. Stud.* **2014**, *38*, 228–237. [CrossRef]
3. Organic Milk Market Report 2017. Available online: http://www.omsco.co.uk/marketreport (accessed on 24 July 2017).
4. Hughner, R.S.; McDonagh, P.; Prothero, A.; Shultz, C.J.; Stanton, J. Who are organic food consumers? A compilation and review of why people purchase organic food. *J. Consum. Behav.* **2007**, *6*, 94–110. [CrossRef]
5. Rosati, A.; Aumaitre, A. Organic dairy farming in europe. *Livest. Prod. Sci.* **2004**, *90*, 41–51. [CrossRef]
6. Bani, P.; Sandrucci, A. Yield and quality of milk produced according to the organic standards. *Sci. Tec. Latt.-Casearia* **2003**, *54*, 267–286.
7. Mueller, C.; de Baan, L.; Koellner, T. Comparing direct land use impacts on biodiversity of conventional and organic milk—Based on a swedish case study. *Int. J. Life Cycle Assess.* **2014**, *19*, 52–68. [CrossRef]
8. Sundberg, T.; Berglund, B.; Rydhmer, L.; Strandberg, E. Fertility, somatic cell count and milk production in swedish organic and conventional dairy herds. *Livest. Sci.* **2009**, *126*, 176–182. [CrossRef]
9. Müller-Lindenlauf, M.; Deittert, C.; Köpke, U. Assessment of environmental effects, animal welfare and milk quality among organic dairy farms. *Livest. Sci.* **2010**, *128*, 140–148. [CrossRef]
10. Stiglbauer, K.E.; Cicconi-Hogan, K.M.; Richert, R.; Schukken, Y.H.; Ruegg, P.L.; Gamroth, M. Assessment of herd management on organic and conventional dairy farms in the united states. *J. Dairy Sci.* **2013**, *96*, 1290–1300. [CrossRef] [PubMed]
11. Schwendel, B.H.; Wester, T.J.; Morel, P.C.; Tavendale, M.H.; Deadman, C.; Shadbolt, N.M.; Otter, D.E. Invited review: Organic and conventionally produced milk-an evaluation of factors influencing milk composition. *J. Dairy Sci.* **2015**, *98*, 721–746. [CrossRef] [PubMed]

12. Capuano, E.; Gravink, R.; Boerrigter-Eenling, R.; van Ruth, S.M. Fatty acid and triglycerides profiling of retail organic, conventional and pasture milk: Implications for health and authenticity. *Int. Dairy J.* **2015**, *42*, 58–63. [CrossRef]

13. Butler, G.; Nielsen, J.H.; Slots, T.; Seal, C.; Eyre, M.D.; Sanderson, R.; Leifert, C. Fatty acid and fat-soluble antioxidant concentrations in milk from high- and low-input conventional and organic systems: Seasonal variation. *J. Sci. Food Agric.* **2008**, *88*, 1431–1441. [CrossRef]

14. Benbrook, C.M.; Butler, G.; Latif, M.A.; Leifert, C.; Davis, D.R. Organic production enhances milk nutritional quality by shifting fatty acid composition: A united states-wide, 18-month study. *PLoS ONE* **2013**, *8*, e82429. [CrossRef] [PubMed]

15. Popovic-Vranjes, A.; Savic, M.; Pejanovic, R.; Jovanovic, S.; Krajinovic, G. The effect of organic milk production on certain milk quality parameters. *Acta Vet.* **2011**, *61*, 415–421. [CrossRef]

16. Kelley, N.S.; Hubbard, N.E.; Erickson, K.L. Conjugated linoleic acid isomers and cancer. *J. Nutr.* **2007**, *137*, 2599–2607. [PubMed]

17. Churruca, I.; Fernández-Quintela, A.; Portillo, M.P. Conjugated linoleic acid isomers: Differences in metabolism and biological effects. *BioFactors* **2009**, *35*, 105–111. [CrossRef] [PubMed]

18. Butler, G.; Nielsen, J.H.; Larsen, M.K.; Rehberger, B.; Stergiadis, S.; Canever, A.; Leifert, C. The effects of dairy management and processing on quality characteristics of milk and dairy products. *NJAS Wagening. J. Life Sci.* **2011**, *58*, 97–102. [CrossRef]

19. Kelsey, J.; Corl, B.A.; Collier, R.J.; Bauman, D.E. The effect of breed, parity, and stage of lactation on conjugated linoleic acid (CLA) in milk fat from dairy cows. *J. Dairy Sci.* **2003**, *86*, 2588–2597. [CrossRef]

20. Elgersma, A.; Tamminga, S.; Ellen, G. Modifying milk composition through forage. *Anim. Feed Sci. Technol.* **2006**, *131*, 207–225. [CrossRef]

21. Collomb, M.; Schmid, A.; Sieber, R.; Wechsler, D.; Ryhänen, E.-L. Conjugated linoleic acids in milk fat: Variation and physiological effects. *Int. Dairy J.* **2006**, *16*, 1347–1361. [CrossRef]

22. Butler, G.; Collomb, M.; Rehberger, B.; Sanderson, R.; Eyre, M.; Leifert, C. Conjugated linoleic acid isomer concentrations in milk from high- and low-input management dairy systems. *J. Sci. Food Agric.* **2009**, *89*, 697–705. [CrossRef]

23. Bergamo, P.; Fedele, E.; Iannibelli, L.; Marzillo, G. Fat-soluble vitamin contents and fatty acid composition in organic and conventional italian dairy products. *Food Chem.* **2003**, *82*, 625–631. [CrossRef]

24. Palupi, E.; Jayanegara, A.; Ploeger, A.; Kahl, J. Comparison of nutritional quality between conventional and organic dairy products: A meta-analysis. *J. Sci. Food Agric.* **2012**, *92*, 2774–2781. [CrossRef] [PubMed]

25. Srednicka-Tober, D.; Baranski, M.; Seal, C.J.; Sanderson, R.; Benbrook, C.; Steinshamn, H.; Gromadzka-Ostrowska, J.; Rembialkowska, E.; Skwarlo-Sonta, K.; Eyre, M.; et al. Higher pufa and *n*-3 pufa, conjugated linoleic acid, alpha-tocopherol and iron, but lower iodine and selenium concentrations in organic milk: A systematic literature review and meta- and redundancy analyses. *Br. J. Nutr.* **2016**, *115*, 1043–1060. [CrossRef] [PubMed]

26. Simopoulos, A.P. The importance of the ratio of omega-6/omega-3 essential fatty acids. *Biomed. Pharmacother.* **2002**, *56*, 365–379. [CrossRef]

27. Manzi, P.; Pizzoferrato, L.; Rubino, R.; Pizzillo, M. Valutazione di composti della frazione insaponificabile del latte vaccino proveniente da diversi tipi di allevamento. In *Atti del 10° Congresso Italiano di Scienza e Tecnologia Degli Alimenti*; Porretta, S., Ed.; Chiriotti: Pinerolo, Italy, 2012; pp. 482–486.

28. Pizzoferrato, L.; Manzi, P.; Marconi, S.; Fedele, V.; Claps, S.; Rubino, R. Degree of antioxidant protection: A parameter to trace the origin and quality of goat's milk and cheese. *J. Dairy Sci.* **2007**, *90*, 4569–4574. [CrossRef] [PubMed]

29. Puppel, K.; Sakowski, T.; Kuczynska, B.; Grodkowski, G.; Golebiewski, M.; Barszczewski, J.; Wrobel, B.; Budzinski, A.; Kapusta, A.; Balcerak, M. Degrees of antioxidant protection: A 2-year study of the bioactive properties of organic milk in poland. *J. Food Sci.* **2017**, *82*, 523–528. [CrossRef] [PubMed]

30. Mogensen, L.; Kristensen, T.; Søegaard, K.; Jensen, S.K.; Sehested, J. Alfa-tocopherol and beta-carotene in roughages and milk in organic dairy herds. *Livest. Sci.* **2012**, *145*, 44–54. [CrossRef]

31. Cashman, K.D. Milk minerals (including trace elements) and bone health. *Int. Dairy J.* **2006**, *16*, 1389–1398. [CrossRef]

32. Poulsen, N.A.; Rybicka, I.; Poulsen, H.D.; Larsen, L.B.; Andersen, K.K.; Larsen, M.K. Seasonal variation in content of riboflavin and major minerals in bulk milk from three danish dairies. *Int. Dairy J.* **2015**, *42*, 6–11. [CrossRef]

33. Chambers, L. Iodine in milk—Implications for nutrition? *Nutr. Bull.* **2015**, *40*, 199–202. [CrossRef]

34. Bath, S.C.; Button, S.; Rayman, M.P. Iodine concentration of organic and conventional milk: Implications for iodine intake. *Br. J. Nutr.* **2012**, *107*, 935–940. [CrossRef] [PubMed]

35. Payling, L.M.; Juniper, D.T.; Drake, C.; Rymer, C.; Givens, D.I. Effect of milk type and processing on iodine concentration of organic and conventional winter milk at retail: Implications for nutrition. *Food Chem.* **2015**, *178*, 327–330. [CrossRef] [PubMed]

36. Rey-Crespo, F.; Miranda, M.; López-Alonso, M. Essential trace and toxic element concentrations in organic and conventional milk in nw spain. *Food Chem. Toxicol.* **2013**, *55*, 513–518. [CrossRef] [PubMed]

37. Givens, D.I.; Lovegrove, J.A. Invited commentary: Higher pufa and *n*-3 pufa, conjugated linoleic acid, alpha-tocopherol and iron, but lower iodine and selenium concentrations in organic milk: A systematic literature review and meta- and redundancy analyses. *Br. J. Nutr.* **2016**, *116*, 1–2. [CrossRef] [PubMed]

38. SINU Società Italiana di Nutrizione Umana. *Larn-Livelli di Assunzione di Riferimento di Nutrienti ed Enregia per la Popolazione Italiana*; SICS: Milan, Italy, 2014.

39. Manzi, P.; di Costanzo, M.G.; Mattera, M. Updating nutritional data and evaluation of technological parameters of italian milk. *Foods* **2013**, *2*, 254–273. [CrossRef] [PubMed]

40. EFSA Panel on Dietetic Products, Nutrition and Allergie. Scientific opinion on dietary reference values for vitamin A. *EFSA J.* **2015**, *13*, 4028. [CrossRef]

41. EFSA Panel on Dietetic Products, Nutrition and Allergie. Scientific opinion on dietary reference values for vitamin E as α-tocopherol. *EFSA J.* **2015**, *13*, 4149. [CrossRef]

42. EFSA Panel on Dietetic Products, Nutrition and Allergie. Scientific opinion on dietary reference values for iodine. *EFSA J.* **2014**, *12*, 3660. [CrossRef]

43. Agabriel, C.; Cornu, A.; Journal, C.; Sibra, C.; Grolier, P.; Martin, B. Tanker milk variability according to farm feeding practices: Vitamins A and E, carotenoids, color, and terpenoids. *J. Dairy Sci.* **2007**, *90*, 4884–4896. [CrossRef] [PubMed]

44. Fiedor, J.; Burda, K. Potential role of carotenoids as antioxidants in human health and disease. *Nutrients* **2014**, *6*, 466–488. [CrossRef] [PubMed]

45. Flachowsky, G.; Franke, K.; Meyer, U.; Leiterer, M.; Schöne, F. Influencing factors on iodine content of cow milk. *Eur. J. Nutr.* **2014**, *53*, 351–365. [CrossRef] [PubMed]

46. Jensen, M. Comparison between conventional and organic agriculture in terms of nutritional quality of food—A critical review. *CAB Rev.* **2013**, *8*, 1–13. [CrossRef]

47. Huber, M.; Rembiałkowska, E.; Średnicka, D.; Bügel, S.; van de Vijver, L.P.L. Organic food and impact on human health: Assessing the status quo and prospects of research. *NJAS Wagening. J. Life Sci.* **2011**, *58*, 103–109. [CrossRef]

48. Baranski, M.; Rempelos, L.; Iversen, P.O.; Leifert, C. Effects of organic food consumption on human health; the jury is still out! *Food Nutr. Res.* **2017**, *61*, 1287333. [CrossRef] [PubMed]

49. Tuomisto, H.L.; Hodge, I.D.; Riordan, P.; Macdonald, D.W. Does organic farming reduce environmental impacts?—A meta-analysis of european research. *J. Environ. Manag.* **2012**, *112*, 309–320. [CrossRef] [PubMed]

MDPI AG

St. Alban-Anlage 66

4052 Basel, Switzerland

Tel. +41 61 683 77 34

Fax +41 61 302 89 18

http://www.mdpi.com

Beverages Editorial Office

E-mail: beverages@mdpi.com

http://www.mdpi.com/journal/beverages

www.ingramcontent.com/pod-product-compliance
Lightning Source LLC
Chambersburg PA
CBHW041218220326
41597CB00033BA/6010